Understanding the Energy Transition

Natalia Magnani • Giovanni Carrosio

Understanding the Energy Transition

Civil society, territory and inequality in Italy

palgrave
macmillan

Natalia Magnani
Department of Sociology and Social Research
University of Trento
Trento, Italy

Giovanni Carrosio
Department of Social and Political Sciences
University of Trieste
Trieste, Italy

ISBN 978-3-030-83480-7 ISBN 978-3-030-83481-4 (eBook)
https://doi.org/10.1007/978-3-030-83481-4

This Palgrave Macmillan imprint is published by the registered company Springer Nature Switzerland AG.
The registered company address is: Gewerbestrasse 11, 6330 Cham, Switzerland

Contents

List of Figures

List of Tables

1

Introduction

Natalia Magnani

Abstract The Introduction highlights that, despite the predominance of engineering and economic sciences in the analysis of the energy and climate issue, a sociological approach to the energy transition appears topical. In fact energy is not only a basic factor, whose diversified forms structure the temporal and spatial organisation of social life, but it is also profoundly influenced by the latter. Human and social behaviours are central to future energy scenarios.

Keywords Sociology of energy • Energy research • Social behaviours • Social habits • Energy transition

The general theme of this book concerns the relationship between energy and society, addressed with the tools of the social sciences, in particular of sociology. In recent times, energy, its consumption and its production have been at the centre of the public and scientific debate, as well as the subject of

N. Magnani (✉)
Department of Sociology and Social Research, University of Trento, Trento, Italy
e-mail: natalia.magnani@unitn.it

N. Magnani, G. Carrosio, *Understanding the Energy Transition*,
https://doi.org/10.1007/978-3-030-83481-4_1

national and European policies, as they are closely connected to highly topical issues, intertwined with each other in complex and often contradictory ways: climate change, the reduction of emissions, natural disasters, the security of supply and the sustainability of the current model of economic development. This growing social relevance of energy also emerges in contemporary sociological reflection. An internationally renowned sociologist like J. Urry in all of his latest writings (2011, 2013, 2014) points out that energy is a massive problem for social theory and for twenty-first-century societies. He highlights how similarly to the issue of climate change, "society is the origin of the problem and the solution" (Urry, 2011, p. 55) on the issue of energy.

Energy is not only a "basic factor" (Schumacher cit. in Urry, 2014, p. 3), whose diversified forms structure the temporal and spatial organisation of social life, but it is also at the same time profoundly influenced by the latter. Human and social behaviours are therefore central to future energy scenarios. However, these issues have until very recently been mainly addressed by the economic and engineering sciences. Sovacool (2014), based on the analysis of several thousand articles published in leading energy journals in a period of 25 years since the 1973 energy crisis, highlighted a series of shortcomings and prejudices that characterise energy research. First, the social elements of energy systems and their consequences are often neglected or underestimated. The research focus is on hardware rather than on the human and social software that underlies energy systems. Among the underestimated social phenomena, there are, for example, the social factors and dynamics underlying the demand for energy services and the use of technologies, social perceptions regarding energy risks, communication methods in relation to energy choices, decision-making processes on energy concerning individuals, organisations and communities and the relationship between ethics and energy.

A second problem identified by the authors is that defined as "disciplinary chauvinism", a position that considers most social sciences as secondary and marginal in favour of physical, engineering and economic sciences. The third prejudice is identified with the term "homogeneous perspective" and refers to the fact that the published literature on the topic of energy is predominantly male and the authors come predominantly from Western countries. This imbalance is then reflected in a clear prevalence of studies concerning the problems of the northern part of the world and less attention to issues such as energy poverty, inequalities in access to energy services and the gender dimension of energy issues.

Similarly, Urry (2011) highlights the limits of the dominant approach to the study of the social components of energy, which he identifies with the economic sciences. In particular, the author emphasises how the view of the economy is based on a model of human behaviour and individualistic change and is focused on maximising profit. However, energy systems are not only economic or technological, but they also presuppose patterns of social life that become embedded and therefore acquire an inertia and resistance for long periods of time (carbon lock-in process), as happened for the system based on steel and oil and centred on the use of the car. These systems consist of an intertwining of economy and society. They also presuppose and at the same time contribute to forming habits and routines that are the core of social life and cannot easily be changed, especially in a situation of low trust in institutions and already large and growing inequalities between the rich and poor in income, access to services and lifestyles. Reaffirming the legitimacy and centrality of a sociological approach to energy appears urgent according to various authors since the more engineering and economic sciences continue to be the main disciplines involved in the analysis of the energy and climate issue, the less likely it will be that policymakers will be able to generate the social preconditions for the transition to a low-emission society.

This book starts from the aforementioned considerations and has the general objective of highlighting the sociological relevance of the energy question. The volume is divided into six chapters. Chapter 2 traces the history and development of sociological research concerning the relationship between energy and society. In particular, it is emphasised that in the past, energy has been an ever-present, albeit marginal and unsystematic theme of the sociology of the environment. Recently, however, we have witnessed a re-emergence of interest in energy issues by the social sciences. The chapter outlines the state of the art and the future prospects of sociological studies on energy. It is structured in three parts: the first will go through the main stages of the relationship between sociology and energy, identifying the historical phases in which most of the studies and the types of approaches are thickened; the second part will identify the main issues faced by contemporary sociology, highlighting as well that in Europe and specifically Italy, there is a tradition of recent studies on energy. The third part deals with studies on energy transition, focusing from a critical perspective on the dominant analytical approaches. Finally, we will identify the main issues around which it is possible to develop future research, with particular attention to the relationship between energy and power.

Chapter 3 deals with the central theme from which the recent sociological debate about the energy transition has developed, namely the social acceptability of renewable energies in their most visible and immediate form: the construction and location of renewable energy production facilities. In this sense, social acceptability has to do strictly with a first form of civil society participation in the energy transition that sees the public as a social actor protagonist of conflicts concerning renewable energy. In the first part of the chapter, we start from the critical deconstruction of the Not In My Backyard (NIMBY) syndrome and its assumptions in order to highlight them and then move to how other analytical constructs are more appropriate to understand sociologically local conflicts concerning renewable energies. These include the concept of environmental justice (hence the analysis of the distributive effects of the benefits and risks of territorial policies), the connected issues of participation and consensus and the dynamics of deterritorialisation and reterritorialisation. In the second part of the chapter, we consider some of the most relevant conflicts over renewable energy sources in Italy, in particular biomass and mini-hydroelectric. We highlight the key narratives that characterise these recently emerged conflicts and the way they challenge NIMBY definitions.

Chapter 4 looks at the energy transition from the point of view of social and territorial inequalities. The energy transition, in fact, takes on various connotations according to the territorial contexts and unequally involves different social groups. In particular, energy transition policies favour or penalise specific places and people. The literature has addressed the issue of inequalities in two ways: first, by investigating conflicts related to the location of renewable energy plants (see Chap. 3); and second, in relation to the energy poverty issue, by highlighting how problems of access to energy as a fundamental good still exist. In this chapter, we address instead the issue of inequalities in a third way: we want to investigate how the policies for the diffusion of renewable energies, for energy saving and energy efficiency, favour or disadvantage weaker social groups and marginal territories, producing or reproducing social and spatial inequalities.

Chapter 5 analyses another aspect of the involvement of civil society in the energy transition. The focus is in fact the emergence from below of new forms of collective action which, similarly to what has happened in other sectors, such as food, unite producers and consumers of renewable energy and intend to give life to new exchange systems. These innovative forms of prosumption can be analysed through the dominant construct of community energy. This is a broad concept with blurred borders encompassing a variety of experiences concerning communities of interest or local communities, which share the development of a project for the production of renewable energy and the economic and social benefits that derive from it. The focus is on the new forms of social and collective relations produced by citizen participation in the energy market.

In spite of community energy being a recent research object, in the first part of the chapter, we will highlight the key theoretical perspectives that have been emerging in the literature. In the second part of the chapter, we will focus on community energy in Italy. Among the different experiences, particular attention will be given to the emergence in Italy of different types of cooperatives involved in the production and distribution of renewable energy. In the conclusion, we will try to give some interpretation of the difficulties faced by community energy initiatives in Italy.

Finally, in Chap. 6, we will readdress the themes analysed in the various chapters, focusing on the role of sociology in the energy transition.

References

Sovacool, B. K. (2014). What are we doing here? Analyzing fifteen years of energy scholarship and proposing a social science research agenda. *Energy Research & Social Science, 1*, 1–29.

Urry, J. (2011). *Climate change and society*. Polity Press.

Urry, J. (2013). *Societies beyond oil*. Zed Books.

Urry, J. (2014). The problem of energy. *Theory, Culture & Society, 31*(5), 3–20.

2

The Emergence of the Sociology of Energy

Giovanni Carrosio

Abstract This chapter outlines the state of the art and the future prospects of sociological studies on energy. It is structured in three parts: the first will go through the main stages of the relationship between sociology and energy, identifying the historical phases in which most of the studies and the types of approaches are thickened; the second part will identify the main issues faced by contemporary sociology, highlighting as well that in Europe there is a tradition of recent studies on energy. The third part deals with studies on energy transition, focusing from a critical perspective on the dominant analytical approaches. Finally, we will identify the main issues around which it is possible to develop the future research.

Keywords Sociology of energy • Theoretical approaches • Power • Disciplinary issues • Transition

---

G. Carrosio (✉)
Department of Social and Political Sciences, University of Trieste, Trieste, Italy
e-mail: gcarrosio@units.it

N. Magnani, G. Carrosio, *Understanding the Energy Transition*,
https://doi.org/10.1007/978-3-030-83481-4_2

2.1 A New Sociological Interest in Energy Issues

For some years, energy has returned to being a central theme in the public debate. The concomitance of some factors, such as climate change, the progressive depletion of fossil resources and the growing precariousness of energy supplies due to the instability of the main supplier countries, means that a large part of the political agenda of the European Union is focused on the energy issue. Europe essentially moves on two levels, that of the environment and that of security. The first requires reducing emissions by replacing fossil sources with renewable ones, saving energy and increasing energy efficiency. On this front, the European Green Deal hinges on ambitious decarbonisation goals. The second involves the construction of new infrastructures to expand the number of supplier countries. In particular, the natural gas sector is at the centre of important strategies: the security of supply is strongly influenced by geographical and political elements. This creates considerable uncertainty in a period in which production within the European Union is rapidly decreasing and supplies from third countries are highly concentrated. On a global level, it is difficult to interpret the situation. According to many observers, we are in the midst of an energy transition from a system based mainly on fossil sources to one centred on renewables (Hicks & Nelder, 2008; RMI, 2014), with many factors of uncertainty regarding the future applications of nuclear power and heavy investments in gas as a transitional vector. However, there are also different positions, which emphasise the new discoveries in the field of shale oil, the oil obtained from clayey shales in which the United States, China and India are investing. If the scenarios traced by some analysts are translated into reality (IEA, 2013; contra Kumhof & Muir, 2012), the new discoveries could represent an important brake on the transition, shifting peak oil by a few decades and putting fossil resources at the heart of energy systems. These important processes, which are changing the balance from the global scale to the local energy systems, are accompanied by a re-emergence of the interest of the social sciences in energy issues (Osti, 2010). In particular, sociology—which historically has never developed a structured interest in energy (Lutzenhiser, 1994), alternating between short periods of rediscovery of the theme with long phases of oblivion—has seen a growth in works that, according to different perspectives, address the energy issue. Compared to how historically the relationship between energy and sociology has evolved, we are in a phase of the greatest flowering of publications and issues being addressed (Holleman, 2012). However, while in the past some lines of research have elaborated in depth the relationship between energy and society to promote a sociology that uses the theoretical foundations of energy to understand society (energy sociology), today the studies focus mostly on the problems related to the production, dispatching and consumption of energy, to be addressed through the traditional toolbox of sociological theories (sociology of energy).

This chapter aims to trace the state of the art of sociological studies on energy. It is structured in two parts: in the first, the main stages of the relationship between sociology and energy are addressed, identifying the historical phases in which most of the studies are concentrated and the types of approaches used; in the second part, the main issues addressed by contemporary sociology will be identified, highlighting how a recent tradition of energy studies also exists in Italy.

2.2 Sociology and Energy: A Historical-disciplinary Path

In the space of this chapter, we cannot reconstruct in detail the relationship between energy and sociology. Other authors have already worked on the *historicisation* of this relationship (Strassoldo, 1983; Rosa et al., 1988). Instead, we will identify the most important historical-disciplinary phases on which the foundations were built and the interests of sociology for the energy issue were concentrated.

Similarly to the traditional distinction made within the sociology of the environment (Dunlap et al., 1979), we can trace a division between two conceptions in the relationship between energy and sociology (Strassoldo, 1983): on the one hand, an *energetic sociology*, which tried to construct a general theory of society starting from the application of the laws of energy to the study of social facts; on the other, a sociology of energy, which instead dealt with the study of the social problems inherent in the production, dispatching, consumption and saving of energy.

The first used energy as a cognitive tool, whereas the second made it an object of study. In the history of sociology, these two conceptions follow particular temporal phases.

Energy approaches were born between the end of the nineteenth century and the beginning of the twentieth century, in a full positivistic climate, to re-emerge between the 1970s and 1980s of the twentieth century with the return of the contamination between sociology and ecology (Rosa & Machlis, 1983). Pioneers of this approach were Ernest Solvay (1906), founder of the Free Institute of Sociology in Brussels, and Wilhelm Ostwald (1909), Nobel laureate in chemistry.

They founded the socio-energetic, a line of studies that had a dual ambition: to interpret the evolution of societies through energetics and to introduce their laws in the study of social organisations in order to make them more efficient. According to these authors, confident in the role of science and technology in improving social organisation, societies can be interpreted through the laws of energy, and their evolution is intrinsically linked to energy intensity and the ability to make consumption more efficient.

According to Ostwald, every social organisation gains legitimacy, thanks to a virtuous relationship between intensity and energy efficiency, and the functioning of the social order is attributable to the laws of thermodynamics. Weber (1909) also intervened in the foundations of socio-energetic, formulating a critique of the postulates of this approach. In his text "For a cultural theory of energy", Weber highlights how the absence of cultural variables in the socio-energetic formulations makes this approach excessively deterministic and unable to distinguish between different types of social organisations. There would therefore be no "energy ethics" at the base of social differentiation but a socio-cultural propensity that leads, in some contexts, to either intensify or not the production of energy or to maximise (or not) the efficiency in social organisation.

Weber's critique marks the failure of this approach within the future events of sociology, which during the twentieth century increasingly emancipated itself from any "sensitivity to environmental, technical and material aspects, and therefore also energy, by rejecting them outside its models of social system" (Strassoldo, 1983, p. 168). General sociology, in an attempt to institutionalise itself and define the boundaries with respect to other disciplines, has increasingly marginalised the environmental, technical and material aspects to follow the Durkheimian premise of explaining social facts through other social facts. However, it has had moderate success in other disciplines: Ostwald and Solvay's energetics, for example, was taken up by the anthropologist White who made it a real theoretical paradigm based on the idea that culture was "a thermodynamic system in the mechanical sense, which develops all its aspects—ideological, technological and sociological—to the extent that the quantity of energy absorbed per capita in a year increases and the means of using this energy are improved" (1969, p. 330).

Between the 1970s and 1980s of the 1900s, with the emergence of contamination between sociology and ecology, and therefore with human ecology, the attempt to build new disciplinary interdependencies resurfaced, starting also with energy. At the international level, the contributions of Machlis (1979) and of Rosa and Machlis (1983) are important, following the pioneering works of Cottrell (1955) and a growing interest of sociology for environmental variables (Catton, 1980; Dunlap et al., 1979). They resume Ostwald's work in an attempt to provide a metatheorical perspective to the renewed interest in energy problems. In Italy, Strassoldo's work is grafted onto this line of studies through a series of contributions (1974, 1976, 1983), which aspired to lay the foundations of the human ecology approach in the Italian country. Beyond an exchange with the environmental economy, which in the same years developed a rich scientific production (Georgescu-Roegen, 1971), these works, as happened for environmental sociology, did not have the desired follow-up due to the widespread tendency to conceive of energetic sociology as one sociological branch among others and the adherence of most scholars to the dominant sociological paradigm. Little remains, therefore, of the socio-energetic in contemporary sociology.

Strassoldo's deterministic approach, the underestimation of cultural variables and the linear conception of social evolution were critical elements for sociological theory. The socio-economic situation of the historical period in which it emerged, characterised by strong economic expansion and a profound optimism towards scientific-technological progress, also contributed to these limits of socio-energetic. Both in the United States and in Europe, socio-energy was born as an intellectual response to the then disruptive processes of industrialisation, urbanisation, the growing division of labour and exploitation of natural resources. The environment was seen as an element of influence with respect to culture and organisational forms, and not vice versa.

The sociology of energy, on the other hand, appeared in the 1940s and 1950s of the twentieth century, with a series of studies on the social and political effects of the diffusion of electronuclear plants (Duncan, 1978), with their promise of unlimited energy, connected risks, localisation needs and problems linked to the oligopoly of nuclear energy (Whitney, 1950; Sylos Labini, 1959). Alongside the nuclear question, which will remain an open field of study for several years and will be addressed according to the national declinations to the problem, in the post-war period, industrialisation led to the massive construction of large dams for the exploitation of water, which were followed by a series of studies devoted to the relationship between energy infrastructures and local communities, dams and social control of water (Smith, 1968).

In recent years, the environmental issue has been gaining ground, and research and models related to the idea of the finiteness of natural resources and the limits of development have begun to spread. Energy is not only an indispensable and limiting condition for economic development, it is also a factor strongly linked to environmental sustainability since "every production process involves a dissipation of heat into the environment and this can lead in the long run to catastrophic phenomena. global level and acute problems at the local level" (Strassoldo, 1983, p. 170).

The 1973 oil crisis, in particular, was an event of central importance because it ignited the sociological debate on the issue of resource scarcity and highlighted for the first time how the environmental issue was just as important as any other social problem. The scarcity of resources has shown that environmental conditions can have direct social consequences. Furthermore, the oil crisis has highlighted how the availability of energy sources does not depend only on a scientific evaluation of resources but on how they are managed and treated in geopolitical equilibrium. Following the oil crisis, there was a first early change in international energy policies. Countries like France decided to invest in nuclear power. Other countries, such as Japan and Sweden, started to consider energy efficiency policies. Nuclear energy became the subject of one of the most protracted and heated technological controversies in modern society.

This controversy is addressed in most of the attempts at sociological analysis of the energy question and has produced a considerable amount of empirical evidence (Freudenburg & Rosa, 1984). Nuclear energy has been analysed as a social movement (Walsh, 1981), as an arena for studying public perception and social acceptance of technological risk (Slovic et al., 1981), as an opportunity to test theories of collective rational action (Walsh & Warland, 1983), as an important example of the difficulty of governing technology in a democratic way (Nelkin, 1977).

The energy question is beginning to join up with the environmental one, and there are many questions about the sustainability of supplies and the relationship between energy technologies and development models. The growing attention to the development and exploitation of domestic energy resources produces new pressures at the level of local communities that are faced with the construction of new plants, related not only to nuclear power plants but also to coal plants, as well as to oil and gas pipelines. Between the 1970s and 1980s, especially in the context of the American environmental sociology, studies on local communities and on the impacts produced by energy policies emerged and multiplied (Rosa et al., 1988). In the 1980s, the environment and energy also became themes of political and ideological mobilisation. Non-academic intellectuals such as Commoner (1976) and Illich (1973, 1983) have addressed the energy issue as a foundation for the construction of a critical thinking towards the dominant technological development. The main field of study is that of energy saving, which has seen a flourishing of research at an international level (Mayer, 1978). The most followed approach in this phase is the psycho-social reading of consumer attitudes and behaviours, but there is no lack of more strictly sociological approaches, such as the study of habits (Palmborg, 1986). The works on energy consumption represent the link and the element of continuity between this phase and the contemporary re-emergence of sociological interest in energy issues.

2.3 The Rediscovery of Energy

In recent years, following the fusion between the environmental and energy issues and the European Union policies for energy saving and incentives for renewable energy, the interest of sociologists in energy has grown steadily. The attention to the issue of energy was catalysed by generalist sociologists such as Urry. Interest in energy issues is mediated by reflection on the social implications of the problem of climate change (Bozzini, 2011).

Urry's (2011) contribution focuses in particular on how contemporary society developed and became dominant in the twentieth century thanks to high levels of greenhouse gas emissions—the high-carbon society. The author links the emergence of the high-carbon society to the consolidation of social practices that presuppose specific patterns of movement and consumption. In particular, the distinction between *neighbourhood lives* and *lives lived beyond a neighbourhood* is central. The first style of life characterised the premodern age and saw the predominance of certain forms of slow mobility, such as pedestrian and bicycle mobility. The scale of the distances travelled fell within a few miles, and many patterns of consumption, family and relationships were localised. Society was organised around small worlds. However, in the twentieth century, the supply and distribution systems that connect production and consumption have become much more extensive in space and time, especially because electricity and the automobile have transformed society. The size of the neighborhood has become much less important.

Even in the era of Covid, the return to the home—possible thanks to a large use of electricity and heat—does not imply a return to the localisation of consumption but is based on an extensive marketing networks of products that arrive at home thanks to an increase in mobility, hitherto dependent on fossil sources. According to Urry, the emergence of five interdependent socio-technical systems of production and consumption, which produced a new energy metabolism, played an important role in affirming the current high-carbon development mode: the development of electricity generation systems and national power grids capable of ensuring that almost all homes in the West are heated, illuminated and populated with electrical-based technological systems; the spread of the petroleum car and the construction of roads that connect most of the places of life and work; the suburbanisation of the territory, focused on private mobility, commuting and the spread of consumer goods that require electricity; the emergence of communication technologies to connect with friends, colleagues, family members with whom it is possible to communicate while being distant; the proliferation of specialised leisure places that require travel both by car and, in some cases, by air transport and which often require the movement of objects, water and food over long distances.

This system, with respect to which we see partial fragility in the face of pandemic events, raises a series of questions that have sparked debate in the social sciences. First of all, the centrality of oil raises questions about the sustainability of this model: problems related to climate-altering emissions, those of a geopolitical nature and the finiteness of fossil resources. To date, no real alternatives to oil have emerged, although there are optimistic assumptions about hydrogen. One way out, which we are practising in the months of the pandemic, could be represented by the replacement of physical mobility with diversified forms of digital communication. As early as the 1990s of the last century, the digital future had appeared as something real. The pandemic is strongly accelerating this perspective: now, in fact, digital lives are replacing mobile lives, reducing the need for travel, especially over long distances. It is still to be understood how this model actually impacts on energy consumption. Certainly there is a shift in energy consumption from mobility to domestic electricity, but it is necessary to understand whether these consumptions will come more and more from renewable sources or from fossil sources. Furthermore, the energy consumption of digitalisation is greater than what one might imagine: the research on vast areas in the coldest regions of the planet to install "new digital factories" made up of servers and data storage devices testifies to this.

Depending on the ability of a digitised society to save energy and accelerate the energy transition, following Urry's analysis there could be two scenarios for the coming decades. The first is a worsening of conflicts related to energy resources. There are limited oil reserves left around the world, and no plan has yet been drawn up to replace them. New digital technologies will not be able to minimise energy costs, and the population will resume its growth upon the end of the pandemic. It is a vision of a neo-medieval future: as in the Middle Ages, there will be limited democracy, a reduced capacity of states to govern, numerous illegal movements of populations across national borders and intense conflicts for resources. A more optimistic scenario is represented by the low-carbon society, which sees an organised reduction in energy consumption, in which actions are planned to address the problem of limited energy stocks and at the same time, an attempt to maintain some benefits and comforts of Western societies. This future requires leveraging innovation in the thinking and practices of scientists and civil society. It is on this second scenario that the works of contemporary social scientists focus.

There are numerous areas of research, many of which can be traced back to the trend in studies on consumption, defined as energy saving, green consumerism and co-provision of energy. This is an interesting novelty because before the 1990s attention was never paid to the qualitative aspects of energy consumption, probably for three reasons:

- energy consumption is impalpable and depersonalised, with little connotation of symbolic meanings (Darby, 2006);
- the energy sector was a monopoly of the state for a long time, and the consumer was unable to choose his own energy supplier;
- historically, energy production was physically and psychologically distant from consumers (Shackely & Green, 2007). This aspect also helped to shape their passive and unreflective attitude (Walker & Cass, 2007).

For these reasons, energy consumerism is still a largely unexplored issue. There are only pioneering works, aimed at investigating issues such as the co-supply of energy (Osti, 2010) and the theoretical aspects of the relationship between producers and consumers in the field of renewable energy sources (Osti, 2008). As we will show in Chap. 4, in the international literature, however, we find a large series of works that emphasise the new proximity between production and consumption, especially in countries where the energy sector has been liberalised for the longest time. Studies on energy communities (see Chap. 4; Rogers et al., 2012; Walker et al., 2010), on consumer cultures and practices (Shove & Warde, 2002), on the relationship between consumer attitudes and policies (Ruostetsaari, 2009) and on co-provision (Sauter & Watson, 2007) are now widespread. It is interesting how the theme of energy consumption has also declined, starting from studies on energy poverty, understood as the impossibility of accessing resources or using them adequately (Pachauri & Spreng, 2004). This theme is very relevant in the literature of Western countries, but still little represented in the Italian context.

Studies on energy communities look at the relationship between consumers and energy producers within restricted territorial contexts. The emphasis is mainly on energy as a vehicle for local development (Khan et al., 2007) and social inclusion (van der Horst, 2008). Consumers are seen as an integral part of a local production and consumption system: they participate in the production of energy both as co-providers and as shareholders of the plants and as local stakeholders. The ideas of a short supply chain and direct relationship between producers and consumers, which are already known in the literature on critical consumption, are returning. Much emphasis is placed on the involvement of social enterprises in green energy circuits (Carrosio, 2007, 2011) and on cooperative economic forms (Devin-Wright, 2005).

The studies on practices look at energy consumption inserted within a chain of actions and habits, technical constraints and cultural models that inform the production and use of energy (Shove, 2010; Spaargaren, 2011). This type of approach has until now been applied above all in the energy-saving sector and in the practices of thermal energy consumption inside homes, with particular attention to the interaction between socio-technical regimes and habitus (Hobson, 2003). However, the available studies place little attention on electricity and the importance of electric distribution networks. However, they represent a very complex socio-technical system (Schewe, 2007) which has an important influence on the ability of consumers to determine their choices.

In addition to issues relating to energy consumption, a relevant area of study concerns local conflicts and the social acceptability of new plants.

The issue of conflicts around renewable energies has aroused much interest: in particular in Italy, where thousands of plants have been built in a few years, local opposition has actually reached significant dimensions.

2.4 The Studies on Energy Transition

The rediscovery of energy by the social sciences has reinvigorated transition studies. In fact, if we look at research works that are placed on a more general dimension, starting from the 1990s a dominant perspective has developed which analyses the transition to sustainability: we refer to the multilevel perspective (MLP). The proponents of this perspective are interested in the mechanisms and processes of change that occur at the sectoral level in individual socio-technical systems. Grin et al. (2010) describes this approach underlining that MLP conceives the transition as an interaction between processes occurring at three levels: innovative practices (niche), structure (regime) and long-term exogenous trends (landscape). Only when these different dynamics occur together in particular ways can the situation be realised in which they can reinforce each other. This represents the condition for the transition to occur.

The transition towards sustainability is in this paradigm the result of the transition from a stable socio-technical configuration to a new one as a consequence of the pressures exerted by factors of an external context. These are profound structural trends (e.g., climate change, cultural changes) which destabilise the dominant socio-technical regime and which open up opportunities for the development of alternative socio-technical systems, defined as "niches". The term niches refers to unusual applications, demonstration programmes and social movements of still limited diffusion. These are sustainability experiments in which participation is widespread and the focus is on social learning processes.

At the heart of MLP is the idea of co-evolution of social systems. They are transformed by bringing about a transition from one system to another. This approach was used to analyse various historical cases of socio-technical system development. Recently, one of the main areas to which the multilevel perspective has been applied is precisely the transformation of energy systems, in particular those of power supply.

According to this approach, renewable energies are conceived as a great social experiment that requires transformations on several levels. As Gross and Mautz (2015) underline, in many European countries the expansion of the niche constituted by renewable sources is in fact closely linked to landscape pressures, such as the emergence of a public debate on the environmental and health risks of nuclear energy and fossil fuels and the emergence of social movements that have attempted to spread alternative energy sources. In some countries, these macro-social developments have produced significant transformations in public opinion regarding conventional energy technologies and their often unpredictable consequences. According to the MLP, what can occur is a growing social support for renewable energy together with the destabilisation of the dominant regime due to tensions or misalignments with respect to new economic and social challenges. If this situation is realised, windows of opportunity can be opened that favour the emergence of radical market innovations, followed by more or less extreme changes at the regime level, for example, the integration of new energy technologies into the existing energy regime or the slow reconfiguration of the energy supply system.

The MLP undoubtedly has the merit of having tried to account for the complexity of the transition processes, highlighting how local practices are influenced by multiple types and levels of structuring. All this makes it easier for some technological trajectories to go in one direction rather than another.

However, MLP has recently come under a lot of criticism. Carrosio and Scotti (2019) highlight that the theory was developed through the retrospective analysis of historical transitions, in which *ad hoc* definition of niche and regime was possible. The current conditions are much more complex, and it is difficult to clearly delineate the boundaries between different analytical levels.

Second, the MLP literature is generally concerned with developing niches of technological innovations in competitive markets. There is often a strong regulatory interest in the strategic and managerial management of economic niches in a top-down governance context. The consideration of the role of social innovation and civil society is limited. In general, we can say that in the MLP there is little attention paid to the roles of the public and to the various forms of demands that are produced and emerge in the energy transition. There is also no analysis of the conflictual and political dimension of these forms of participation. As Pellizzoni (2018) points out, in this approach a depoliticised vision of innovation prevails, which is supposed to develop in a substantially unchanged social and economic framework.

Finally, Carrosio and Scotti (2019) believe that this frame underestimates the role of the socio-territorial aspects involved in the energy transition process, which appear to influence the transition paths (both in terms of promoting and resistance to the socio-technical change) and their results at the local level.

Starting from a critique of MLP, various authors highlight the need for a multidimensional conceptualisation of the roles, forms of involvement and potential of the public in the energy system (Walker & Cass, 2007; Chilvers & Longhurst, 2016). A useful contribution in this direction comes from the "Science and Technology Studies" (STS) approaches, according to which technologies are not only material objects but also embedded components of socio-technical systems in which producers, infrastructures, users, consumers, regulators and other intermediaries are all harnessed (Bijker & Law, 1992). According to this perspective, the technical and the social are co-constituted, interact constantly and shape each other with exchanges in both directions. A similar reasoning is proposed by authors who come from the geography and sociology of the environment (Van Vliet et al., 2005), who, however, question themselves above all on the organisational and social forms of the energy transition. For example, Walker and Cass (2007) identify four different socio-technical combinations used to organise energy production: the public utility model, the private supplier model, the family model and the community model.

The universal model of public utilities, characterised by large plants managed by public operators, has largely been superseded (Spaargaren, 2011). Over time, the private sector has become predominant. With this passage, the role of consumers has changed; from passive actors holding the right to the supply of electricity, they have become active actors with the possibility of choosing between different suppliers and different tariffs. In this transformation, the family model has been the focus of rhetorical and material attention on the part of public and private producers.

Furthermore, the spread of micro-devices for energy production and saving has led to the family model evolving. The figure of the *prosumer* has emerged, that is the consumer who also becomes a producer: a social figure that responds to the logic of responsibility, efficiency, independence and self-sufficiency.

A minority role, albeit a growing one, has finally been assumed by the community model that has developed above all in rural areas since the 2000s. These are new forms of non-oppositional collective action in which people are considered above all as a community, that is as participants in or beneficiaries of community projects, able to collectively contribute to various forms of local solidarity. Analysing the energy transition through these different consumer roles means going beyond attention to the production/supply of new technologies for energy production and analysing the role of demand and the behaviour of citizens and inhabitants.

2.5 Conclusion

It is interesting to note that in the face of a flourishing of studies on the socio-environmental problems of renewables and on the energy transition, there is no work on the disputes pending around fossil fuels: think, for example, of Enel's investments in clean coal, which have provided for the conversion to coal from several sites in Italy and which generated opposition from local communities. On closer inspection, this disinterest is generalised to the whole traditional energy system. There are very few works on the role of utilities, on the strategies of resistance to change by the large energy groups, on the orientations of the energy companies' organisations, on the innovations within the technological path of non-renewable sources (Carrosio & Magnani, 2020). These are all elements on which it would be necessary to shed light in order to understand how effectively the energy system is in transition and if and by whom the transition is governed. And again, there is a lot of emphasis on electricity and little interest in heat. Thematising the conflict between production and saving of thermal energy could perhaps highlight some internal problems of the energy transition, recalibrating attention also to the big energy players: in this phase, the traditional energy system is investing in the technological modernisation of its plants to make them more efficient and clean (storage of CO2, use of thermal waste from gas plants, modernisation of turbines, etc.). Investments in efficiency, such as the construction of district heating networks in the large cities of northern Italy, can paradoxically represent a brake on the transition, extending the competitiveness of traditional power plants for many years. Saving energy, on the other hand, implies a reduction in consumption and at the same time a decrease in the turnover of large groups. Starting from this tension, much of the energy controversy, which revolves around the following antinomies, can be resolved: efficiency/savings, centralised generation/widespread generation and safety/autonomy. These are antinomies that also have a strong political significance, which is demonstrated by the re-emergence of studies on the relationship between energy, power and democracy, which highlight the relationship between energy systems and social models and of which Adams is considered a precursor (1975). Mitchell's most recent work (2011) on the relationship between democracy and fossil fuels is accompanied by research on the social consequences of the transition from centralised to widespread production systems. The debate has only just begun and is mostly conducted on a theoretical level, but two well-defined trends are already being outlined, in which we will probably see protagonists in the years to come: some see in widespread micro-generation an element of empowerment of local communities, where control of local energy production and consumption becomes a form of social autonomy from the market and from central powers (Seyfang & Haxeltine, 2012); others highlight the ambiguous nature of the new devices, which would give rise to a surveillance network (microphysics of power) capable of penetrating into families and guiding their behaviour (Hargreaves, 2010). The task of sociology in the coming years will be to give empirical concreteness to these visions, trying to outline which society awaits us if the energy transition towards a carbon-free system will actually have any chance of success.

References

Adams, R. N. (1975). *Energy and structure: A theory of social power*. University of Texas Press.

Bijker, W. E., & Law, J. (1992). *Shaping technology/building society*. MIT Press.

Bozzini, E. (2011). *La sfida del cambiamento climatico*. Istituzioni, lobby e movimenti sociali in Europa, Carocci.

Carrosio, G. (2007). Imprese sociali verdi in aree fragili. *Impresa Sociale, 4*, 129–146.

Carrosio, G. (2011). *I biocarburanti*. Globalizzazione e politiche territoriali, Carocci.

Carrosio, G., & Magnani, N. (2020). District heating and ambivalent energy transition paths in urban and rural contexts. *Journal of Environmental Policy & Planning, 22*, 460–472.

Carrosio, G., & Scotti, I. (2019). The 'patchy' spread of renewables: A socio-territorial perspective on the energy transition process. *Energy Policy, 629*, 684–692.

Catton, W. (1980). *Overshoot*. University of Illinois Press.

Chilvers, J., & Longhurst, N. (2016). Participation in transition(s): Reconceiving public engagements in energy transitions as co-produced, emergent and diverse. *Journal of Environmental Policy and Planning, 18*(5), 585–607.

Commoner, B. (1976). *The poverty of power. Energy and the economic crisis*. Alfred A. Knopf.

Cottrell, F. (1955). *Energy and society*. McGraw-Hill.

Darby, S. (2006). *The effectiveness of feedback on energy consumption*. Environmental Change Institute.

Devin-Wright, P. (2005). Local aspects of UK renewable energy development: Exploring public beliefs and policy implications. *Local Environment: The International Journal of Justice and Sustainability, X*(1), 57–69.

Duncan, O. D. (1978). Sociologists should reconsider nuclear energy. *Social Forces, 57*(1), 1–23.

Dunlap, R., William, R., & Catton, J. R. (1979). Environmental sociology. *Annual Review of Sociology, 5*, 243–273.

Freudenburg, W. R., & Rosa, E. A. (Eds.). (1984). *Public reactions to nuclear power: Are there critical masses?* Westview Press, Inc.

Georgescu-Roegen, N. (1971). *The entropy law and the economic process*. Harvard University Press.

Grin, J., Rotmans, J., & Schot, J. (2010). *Transitions to sustainable development: New directions in the study of long-term transformative change*. Routledge.
Gross, M., & Mautz, R. (2015). *Renewable energies*. Routledge.
Hargreaves, T. (2010). *Putting Foucault to work on the environment: Exploring pro-environmental behaviour change as a form of discipline*. CSERGE Working Paper EDM, 10–11.
Hicks, B., & Nelder, C. (2008). *Profit from the peak. The end of oil and the greatest investment event of the century*. John Wiley & Sons Inc.
Hobson, K. (2003). Thinking Habits into action: The role of knowledge and process in questioning household consumption practices. *Local Environment: The International Journal of Justice and Sustainability, VIII*(1), 95–112.
Holleman, H. A. (2012). *Energy justice and foundations for a sustainable sociology of energy*. Department of Sociology, University of Oregon.
IEA. (2013). *World energy outlook 2013*. International Energy Agency.
Illich, I. (1973). *Energy and equity*. Marion Boyars.
Illich, I. (1983). The social construction of energy, Città del Messico, El Colegio de Mexico.
Khan, M. I., Chhetri, A. B., & Islam, M. R. (2007). Community based energy model: A model approach to developing sustainable energy. *Energy Sources, II*(4), 353–370.
Kumhof, M., & Muir, D. (2012). *Oil and the world economy: Some possible futures*. IMF Working Paper, 12/156.
Lutzenhiser, L. (1994). Sociology, energy and interdisciplinary environmental science. *The American Sociologist, XXV*(1), 58–79.
Machlis, E. G. (1979). *Energy flow and social order: The influence of energy interdependence upon social organization*, PhD Dissertation, Yale University.
Mayer, R. N. (1978). The sociology of consumption: Deriving new social indicators. *Social Indicators Research, V*(1–4), 151–167.
Mitchell, T. (2011). *Carbon democracy: Political power in the age of oil*. Verso Books.
Nelkin, D. (1977). *Technological decisions and democracy*. Sage.
Osti, G. (2008). Relazioni fra consumatori e produttori nel campo delle fonti energetiche rinnovabili. *Sociologia Urbana e Rurale, XXX*(85), 41–56.
Osti, G. (2010). Energia e società: alcuni elementi di base. *Quaderni di Teoria Sociale, 10*, 105–129.
Ostwald, W. (1909). *Energetische Grundlagen der Kulturwissenschaft*. Klinkhardt.
Pachauri, S., & Spreng, D. (2004). Energy use and energy access in relation to poverty. *Economic and Political Weekly, XXXIX*(3), 271–278.

Palmborg, C. (1986). Social habits and energy consumption in single-family homes. *Energy, XI*(7), 643–650.

Pellizzoni, L. (2018). Energia di comunità. Una ricognizione critica della letteratura. In G. Osti, & L. Pellizzoni (a cura di), *Energia e innovazione tra flussi globali e circuiti locali* (pp. 17–41). Edizioni Università di Trieste.

RMI. (2014). *The economics of grid defection. When and where distributed solar generation plus storage competes with traditional utility service*. Rocky Mountain Institute.

Rogers, J. C., Simmons, E. A., Convery, I., & Weatherall, A. (2012). Social impacts of community renewable energy projects: Findings from a woodfuel case study. *Energy Policy, 42*, 239–247.

Rosa, E. A., & Machlis, G. E. (1983). Energetic theories of society: An evaluation review. *Sociological Inquiry, LIII*(2–3), 152–177.

Rosa, E. A., Machlis, G. E., & Keating, K. M. (1988). Energy and society. *Annual Review of Sociology, XIV*, 149–172.

Ruostetsaari, I. (2009). Governance and political consumerism in Finnish energy policy making. *Energy Policy, XXXVII*(1), 102–110.

Sauter, R., & Watson, J. (2007). Strategies for the deployment of microgeneration: Implications for social acceptance. *Energy Policy, XXXV*(5), 2270–2279.

Schewe, P. (2007). *The grid: A journey through the heart of our electrified world*. Joseph Henry Press.

Seyfang, G., & Haxeltine, A. (2012). Growing grassroots innovations: Exploring the role of community-based initiatives in governing sustainable energy transitions. *Environmental Planning, 30*, 381–400.

Shackley, S., & Green, K. (2007). A conceptual framework for exploring transitions to decarbonised energy systems in the United Kingdom. *Energy, XXXII*(3), 221–236.

Shove, E. (2010). Beyond the ABC: Climate change policy and theories of social change. *Environment and Planning A, 42*(6), 1273–1285.

Shove, E., & Warde, A. (2002). Inconspicuous consumption: The sociology of consumption, lifestyles and the environment. In R. Dunlap, F. Buttel, P. Dickens, & A. Gijswijt (a cura di). *Sociological theory & the environment: Classical foundations, contemporary insights*. Rowman and Littlefield.

Slovic, P., Fischhoff, B., & Lichtenstein, S. (1981). Perception and acceptability of risk from energy systems. In A. Baum & J. E. Singer (Eds.), *Advances in environmental psychology, Vol. 3 Energy: Psychological*. Erlbaum.

Smith, C. L. (1968). *The Salt river project of Arizona: Its organization and integration with the community*. Department of Anthropology, University of Arizona.

Solvay, E. (1906). *Formules d'introduction à l'energétique physio et psyco sociologique*. Giard&Brieres.
Spaargaren, G. (2011). Theories of practices: Agency, technology, and culture. Exploring the relevance of practice theories for the governance of sustainable consumption practices in the new world-order. *Global Environmental Change, XXI*(3), 813–822.
Strassoldo, R. (1974). Ambiente, energia e potere: appunti di eco-sociologia. Quaderni dell'Isig, Gorizia.
Strassoldo, R. (1976). Energia. Nuovo dizionario di sociología. Edizioni Paoline.
Strassoldo, R. (1983). Energia e società. *Studi di Sociologia, XXI*(2), 160–180.
Sylos, L. P. (1959). Problemi dell'energia nell'economia italiana. *Rivista Internazionale di Scienze Sociali, XXX*(67), 24–43.
Urry, J. (2011). *Climate change and society*. Polity Press.
van der Horst, D. (2008). Social enterprise and renewable energy: Emerging initiatives and communities of practice. *Social Enterprise Journal, IV*(3), 171–185.
Van Vliet, B., Chappells, H., & Shove, E. (2005). *Infrastructures of consumption. Environmental restructuring of the utility industries*. Earthscan.
Walker, G., & Cass, N. (2007). Carbon reduction, "the public", and renewable energy: Engaging with socio-technical configurations. *Area, XXXIX*, 458–469.
Walker, G., Devine-Wright, P., Hunter, S., High, H., & Evans, B. (2010). Trust and community: Exploring the meanings, contexts and dynamics of community renewable energy. *Energy Policy, XXXVIII*(6), 2655–2663.
Walsh, E., & Warland, R. (1983). Social movement involvement in the wake of a nuclear accident: Activists and free riders in the TMI Area. *American Sociological Review, 48*, 764.
Walsh, E. J. (1981). Resource mobilization and citizen protest in communities around Three Mile Island. *Social Problems, 29*(1), 1–21.
Weber, M. (1909). Energetische kulturtheorien. *Archiv für Sozialwissenschaft und Sozialpolitik, 29*, 575–598.
White, L. (1969). *Energia e civiltà*. Sansoni.
Whitney, V. H. (1950). Comment on some social consequences of atomic power. *American Sociological Review, 15*, 556–557.

3

Civil Society and Conflicts Over Renewable Energies Beyond the NIMBY Syndrome

Natalia Magnani

Abstract This chapter deals with the social acceptance of renewable energies in their most visible and immediate form: the location of production facilities. We start from the critical deconstruction of the Not In My Backyard (NIMBY) syndrome and its assumptions as an analytical construct to understand environmental conflicts over renewables. In the first part of the chapter, we then highlight the added value of other analytical constructs, like the concept of environmental justice, the related issues of participation and consensus and the dynamics of territorialisation. In the second part of the chapter, we consider the most important conflicts over renewable energy sources in Italy, and we highlight the key narratives that characterise them and the way they challenge NIMBY definitions.

Keywords Social acceptance • NIMBY • Environmental conflict • Environmental justice • Territorialisation

---

N. Magnani (✉)
Department of Sociology and Social Research,
University of Trento, Trento, Italy
e-mail: natalia.magnani@unitn.it

N. Magnani, G. Carrosio, *Understanding the Energy Transition*,
https://doi.org/10.1007/978-3-030-83481-4_3

3.1 The Social Acceptance of Renewable Energy Sources

A central theme from which the recent sociological debate on the energy transition developed is that of social acceptance of technologies for the production of energy from renewable energy sources (res).

As pointed out by Wüstenhagen et al. (2007), the literature began to address the social acceptability of renewable energies starting in the 1980s, when various surveys concerning the wind sector were conducted in the United States and Northern Europe. However, this issue remained highly marginal in the international literature until the early 2000s due to the relatively low degree of diffusion of plants at the local level, combined with the general and broad public support for renewable technologies. Recently an increasingly binding international and European Union legislation (e.g. Kyoto Protocol, 1998; Directive on Renewable Energy, 2009) and national and European incentives for the production of renewable energy have generated new attention to the problem.

A widespread recognition of the need for empirical sociological research on specific projects for the exploitation of res in order to better understand the social variables that influence their implementation has emerged. Wüstenhagen et al. (2007) distinguishes three dimensions of the social desirability of res: a socio-political acceptability, a market acceptability and a relationship to local communities. Socio-political acceptability is defined as "wider social acceptability, predisposition by public opinion, stakeholders and policy-makers in relation to new technologies and policies" (Wüstenhagen et al., 2007, p. 2684). This dimension takes the form of a definition of national objectives and policies aimed at encouraging the production of renewable energy in the area through memoranda of understanding, agreements between governments, communication spaces and facilitations for stakeholders.

Market acceptance, on the other hand, concerns the adoption of new technologies by the market and involves investors, companies and even consumers. It is built through incentives, agreements and procedural facilitations. These are instruments mainly programmed at a national level. On the contrary, the location of the plants and the authorisation to build them is mainly a matter of regional competence. Finally, community acceptance concerns the specific acceptance of localisation decisions and renewable energy projects by local stakeholders.

The sociological literature has largely focused on the latter and on the conflicts surrounding the location of renewable energy plants (e.g. Bell et al., 2005, 2013; Breukers & Wolsink, 2007; Jobert et al., 2007; Nadaï, 2007; Wolsink, 2007; Devine-Wright, 2009; Firestone et al., 2009; Pasqualetti, 2011). Understood in this sense, social acceptability has to do with the role of the public as a social actor protagonist in conflicts over renewable sources.

Although energy has always been a source of strong transformation for a territory, as happens for example in relation to urbanisation processes, the emergence of the issue of conflicts over res is the indirect result of national support policies, in particular the introduction of incentive tariffs in European countries in the 2000s.

In the context of the liberalisation of the electricity markets, where the energy arena has become more complex and diversified, these policies have largely been interpreted as incentives for producers and operators to minimise costs and maximise the energy efficiency of the existing technology. The easiest way to reduce costs was to build plants that produce energy on a large scale (Gross & Mautz, 2015, p. 106): this is the case, for example, of photovoltaic plants on the ground consisting of hundreds or thousands of panels, or the case of huge geothermal, wind or offshore wind farms.

According to Gross and Mautz (2015, pp. 106–107) it is possible to distinguish two types of conflicts that the initial and dominant type of response to the stimulation of energy from renewable sources has generated. The former can be described as a local/regional conflict. It mainly concerns the competition between interests concerning the exploitation of specific areas of territory (onshore or offshore). An example is the conflict between the development of offshore wind farms and the coastal tourism industry.

Another type of conflict can be identified within the same environmental movement, especially in those countries of Northern Europe where the strong presence of local environmental organisations has been combined with a wide spread of renewables. This is a conflict that opposes on the one hand the ecological modernisation of the energy sector aimed at protecting the environment and the climate and on the other the conservation and protection of biodiversity. Generally, renewables plants are considered as a technology that comes into conflict with nature—in fact, they often have an impact on the environment by disturbing animal species or negatively modifying the landscape and the ecosystem, thus having environmental costs. These costs must be balanced with the benefits that renewable energies produce above all in terms of lower emissions.

However, this is not a simple task since the research results and the evidence generated by the risk assessment are subject to different interpretations, and this has led organisations and individuals to define different priorities in relation to the same energy technologies. These tensions exert pressure on planners and businesses, intervening specifically to limit the possible locations for renewable energy plants or the dimension of the plants.

In this way, as argued by Gross and Mautz (2015, p. 108), the local conflicts that occur around res are also part of a social learning process in which the opportunities and the limits of a socially acceptable expansion of renewable resources become increasingly evident. It is above all the experiences and learning processes of that part of the population worried about the negative impacts of renewable technologies that can contribute to the debate on the conditions that influence not only environmental sustainability but also social sustainability of the energy transition at the local level.

3.2 The Critique of the NIMBY Syndrome

Most studies analysing conflicts related to renewable energy plants in industrialised countries share the critique of explanations based on the NIMBY syndrome (e.g. Bell et al., 2005; Van der Horst, 2007; Wolsink, 2007; Wüstenhagen et al., 2007; Aitken, 2010; Devine-Wright & Wiersma, 2013). On the one hand, the critique of the NIMBY syndrome characterises a specific phase of the debates over res over the last 30 years. On the other hand, the critical deconstruction of the NIMBY syndrome as *an analytical tool* of the relationships between society and nature is a classic theme of the sociology of the environment (Freudenburg & Pastor,

1992; Luloff et al., 1998; Burningham, 2000). The acronym NIMBY—not in my backyard—emerged between the 1970s and 1980s of the 1900s in relation to the phenomenon of opposition to the location of nuclear plants and emissions-intensive industries, which became a widespread practice initially in the United States and then also in Europe. The press and representatives of institutions and businesses resorted to the development of a short and effective label to identify these citizens.

This label, which is often associated with the term syndrome to further underline its pathological aspect, refers to some characteristics that are attributed to the opposition committees. As pointed out by Freudenburg and Pastor (1992) and by more recent empirical research (e.g. Burningham et al., 2015), a variety of visions of the public are associated with the term NIMBY, identifiable in three main attitudes: selfishness, attention only to one's own particularistic and localistic interests; irrationality and excess of emotion, lack of balance and weighting in risk assessment, which leads to the prevalence of prudential attitudes; and superficiality and ignorance of the actual consequences of the proposed initiatives.

These attitudes are usually related to a series of possible consequences: the negation of the common good and of the advantages for the community; the limitation of opportunities for economic development; the affirmation of forms of technological Luddism, in which a few deny the technological progress of many; and a growing mistrust of the main institutional and economic actors.

The use of the NIMBY label has become a tool to question the legitimacy of the actors who constitute local oppositions to the plants. However, through empirical research, both qualitative and quantitative, various studies have highlighted the substantial groundlessness of these accusations to the point of speaking of a real myth of the NIMBY syndrome (Wolsink, 2006).

In particular, as regards the accusation of selfishness, the action of the opposition committees is often opposed not so much to the location of the plant as to the mode of intervention or the type of solution proposed to solve the problem. In addition, it has been shown that the backyard concept is based on an impoverished vision of "home" (Devine-Wright, 2011, p. 62). In fact, it usually refers to a private land. However, the concerns that inflame opponents of plants often transcend individual property boundaries, affecting larger spatial areas and public interest. Furthermore, in many cases the action of the committees has moved beyond the limit of the close localisation battle, forming alliances with other local committees and with supralocal associations.

In general, the overcoming of the localist perspective by the committees is indicated with the acronym Not In Anyone's Backyard (NIABY), an acronym for "not in anyone's backyard," which stresses that the struggle is not against the localisation of the facility in its own courtyard but is against the implementation of that technological solution wherever its location might be.

As regards the issue of emotionality, excess of concern and overestimation of risks not supported by rational arguments, empirical research has shown that in many cases it is precisely the local committees that give evidence of a reasonable evaluation of the risks involved in the construction of the envisaged facility, while the proponents and their experts, for reasons of convenience, underestimate the damages that may result (Pellizzoni, 2011).

In this regard, the literature (Upreti, 2004; Wolsink, 2006) has underlined how the a priori opposition to "other" solutions often characterises the attitudes of the proponents that could also be identified with a label: There Is Not Alternative (TINA) (there is no alternative).

As regards the issue of ignorance and superficiality of the local population, the question of the interaction and integration of expert and profane knowledge with regard to environmental issues assumes importance (Wynne, 1996; Pellizzoni, 2011; Kasperson & Ram, 2013). Although the inhabitants of a place have limited knowledge of the specific functioning of certain technological devices, on the other hand they have a "profane knowledge", a knowledge of the context conditions and an experience of the changes that occur in the local environment which is much more in depth than those—abstract and detached—normally held by experts.

Furthermore, the NIMBY approach tends to hide the question of the post-normality of science in modern society (Funtowicz & Ravetz, 1993). This applies both to the controversial nature of many scientific solutions and to the increasing politicisation of expert knowledge.

These aspects emerge clearly when the local committees themselves resort to counter-expertise to support a different technological and organisational solution. As proof of the non-superficiality of the action of the committees, the result of this widening of the debate is often the raising of the threshold of attention to environmental risks and related controls (Pellizzoni, 2011). Finally, local opposition movements are not necessarily producers of distrust but can instead favour the strengthening of the social bonds of proximity and the mobilisation of some forms of social capital.

Sociological studies on the social acceptance of renewable energies, therefore, start from this critical analysis of the NIMBY syndrome to underline how the use of this term must be considered part of the phenomenon to be investigated. Its use as an analytical tool is instead to be rejected, as it ends up obscuring the real motivations underlying the conflicts and prevents the understanding of what really happens in relation to the location of the renewable energy infrastructures (Wolsink, 2006, 2007).

On a theoretical level, criticising the NIMBY syndrome also means questioning the model of the rational actor who assumes that the social actor acts as *homo oeconomicus* and that human behaviour is fundamentally motivated by personal interest. As we will see in the next paragraphs, the analysis of the conflicts on res instead underlines the relevance of a variety of social and cultural factors concerning the interaction with political and social institutions (justice, trust, territorial attachment, etc.). Implicit in these studies is a questioning of the idea that the opposition to renewable facilities is necessarily negative and instead a demonstration that controversies have a social value in signalling socio-technical problems that may have been overlooked.

3.3 Environmental Justice and Renewable Energies

First of all, the critical analysis of the NIMBY label is also related to the relevance of environmental justice and equity issues. This appears evident from the first North American studies on environmental racism, which highlight marginalisation mechanisms, in which highly polluting activities were located in areas characterised by ethnic minorities and social marginality (Wolsink, 2006).

Research on local conflicts regarding renewable energies draws heavily on this literature and underlines how sentiments of justice and equity, rather than selfishness, are among the determining factors, if not the determining factors, of opposition to facilities (Wolsink, 2007; Kasperson & Ram, 2013).

The literature on environmental justice (e.g. Jenkins et al., 2016) highlights how it consists of two different, but closely interconnected, types of concerns, namely those relating to distributive justice on the one hand and those relating to procedural justice on the other. The term distributive justice refers to the way in which the costs and benefits (economic as well as environmental) associated with a given infrastructure are spatially and socially distributed (Gross, 2007, p. 2729).

Distributive justice therefore has to do with assessing the effects of res developments for local communities. In this regard, empirical research (Upreti, 2004; Jobert et al., 2007; Walker & Devine-Wright, 2008) highlights that the main causes of conflict around renewable energy production infrastructures are the (perceived) reduced benefits for the general local community with respect to the high social and environmental costs and the predominance of interests and profits for a small elite.

Among the costs of the technology for renewable energy production that emerge from the conflict analysis, there are typically the following: health risks—in particular emissions produced by the plant or by the traffic generated to power it, for example in the case of biomass; threats to the safety of the whole community, such as the seismic risk in the case of geothermal energy (Gross & Mautz, 2015, p. 107); problems related to noise pollution (caused, e.g. by wind turbines); the visual impact and landscape transformation (for biogas but also large photovoltaics and wind); and more generally impacts on the quality of life.

Distributive justice concerns not only the right distribution of costs and benefits of a system between social groups but also between territories. Renewable energy plants are often located in rural areas because of the wealth of natural resources (earth, water, and wood). As highlighted by Osti (2013), the demographic and social vulnerability of rural areas, that is the presence of a small and elderly population, exposes them more easily to the risk of colonisation by external economic powers and actors who see a chance for profit in the green economy but do not create opportunities for local development.

As demonstrated by Jobert et al. (2007), the developers of wind farms often come from outside the community, and the general trend is to enter into lease contracts of high economic value with a small number of private owners. In most cases, the local population is not given the opportunity to participate in the design and/or ownership of the wind farm. This further exacerbates the gap between a "few winners" and "many losers" (Gross, 2007). Others (Magnani, 2012a) have shown that the promoters of renewable energy technologies are sometimes powerful local production elites, already discredited as regards their contribution to environmental sustainability, whose commitment to the common interest is therefore considered to be unreliable by most of the local population, based on past experience.

In addition to concerns relating to issues of distributive justice, the literature on res conflicts has also underlined the relevance of procedural justice. It concerns the presence in the territorial planning of a participatory decision-making process capable of guaranteeing to all relevant stakeholders the information and possibility of expressing their different opinions regarding the definition of the problem and solutions (Gross, 2007; Pasqualetti, 2011). Empirical research on the construction of wind farms in various communities in Northern Europe shows that local acceptance of renewable energy depends fundamentally on the transparency of the information provided from the beginning of the decision-making process and on the degree of public participation (Wolsink, 2007). However, the effort of developers to involve the local population in planning and development is often insufficient and of poor quality (e.g. Devine-Wright, 2005; Wolsink, 2007). In some cases, energy infrastructures are built without any information being provided in advance to the communities concerned (Jobert et al., 2007, p. 2759). Often public involvement is limited to information meetings, and the opportunities given to local oppositions to discuss their concerns regarding the structures are generally very limited.

In particular, European research on biomass plants has also reached similar conclusions. Upreti (2004, p. 787), in an analysis of the UK case, highlights that the involvement of the public in the development of biomass energy plants is usually limited to providing economic and technical information on decisions already made. According to the author, the democratic deficit regarding the spread of renewable energy plants reflects the dominant perception of public participation as something that causes delay, underlines the supremacy of technicians, undermines the authority of developers and creates dangerous precedents for future localisation processes (ibid.).

Similarly, other empirical research (Aitken, 2010; Magnani, 2012b) has highlighted the predominance of a technocratic approach to risk in the planning of plants. A top-down risk communication strategy is documented, which dismisses as unfounded public perceptions of risk that differ from the rational explanations of scientists and technicians recruited by the developers. Especially in conflicts on biomass plants, a controversy often emerges between the promoters who affirm the incontrovertible environmental advantages of the proposed technological solution and the local community concerned about the health risks related to emissions and the impact on local ecosystems and tourism (Upreti, 2004, p. 787).

The theme of justice is closely connected to that of trust. As pointed out by Bellaby (2010), in general a great amount of trust is needed to manage the uncertainties and face the risks associated with the energy transition. In the literature that deals with the diffusion of renewable sources, trust is strongly connected to the degree of satisfaction of the stakeholders with the adequacy of the methods of operation of the institutions responsible for the design, implementation, management and regulation of technology (Rayner, 2010; Bellaby, 2010; Kasperson & Ram, 2013). For example, some studies (Bell et al., 2005, 2013) have highlighted how the introduction of economic incentives aimed at altering the public perception of the distribution between costs and benefits of a technology can increase rather than decrease the local opposition if it undermines systemic trust by being seen as a form of corruption.

Furthermore, Gross and Mautz (2015, p. 107) underline the complex relationship between risk management and trust. In the case of geothermal energy, but similarly also for other res, the visible presence of a massive infrastructure, together with the strong dependence on an abstract expert knowledge and the perception of a lack of influence by the population on the decisions concerning a given energy source, can easily help to erode systemic trust.

3.4 Conflicts and Planning of Renewable Energies: Beyond Consensus

On the basis of what has been said so far, part of the literature on renewable energy has developed around the theme of the various ways of involving the public (public engagement) (e.g. Chilvers et al., 2018; Pallett et al., 2019). Haggett (2011) in particular identifies three ideal typical models of public engagement.

Information provision is considered the minimum level of involvement. It is rarely effective in encouraging local communities to support a renewable energy facility; however, it is the most used form by governments and private individuals. The basis of this approach is the belief on the part of those who promote an intervention, especially of the institutions, that the opposition can be tackled simply by closing an information gap through education. It is what in the studies on science and technology (STS) is called a "deficit model" (Wynne, 1996): the idea is that if people had better knowledge, they would support that particular technological development, the renewable energy facility in this case. However, the literature (Kasperson & Ram, 2013) stresses that there is not necessarily a direct correlation between information and behaviour, nor is it sufficient to say that people who oppose a certain work are ill-informed. On the contrary, they are often very knowledgeable about the details of a particular project. Furthermore, attempts to convince citizens of the need for renewable production facilities on the basis of the threat of climate change do not work because local, tangible and immediate problems generally produce more attention than global benefits that are not visible and indirect.

An alternative to offering information to an audience considered passive is to actively provoke responses through involvement in the form of *consultation*. The goal is to create a dialogue between the promoters of the facility and the local population instead of simply activating a one-way flow of information. This requires recognising and giving value to local expertise and the forms of tacit knowledge that different social groups can offer. It also means considering local contingencies and visions of the place. This modality can allow for the elaboration of a "qualified support" for res (Haggett, 2011). The consultation offers the opportunity to discuss with people the topics to which their contribution is important, such as the size of the plant or the impact on fauna and landscape. Arnstein (1969) stresses how this could be a first step towards participation, but much depends on the questions that are allowed and the options that are presented.

A third modality of public involvement concerns the *deliberation*. In this case, the public is not only allowed to discuss possible projects, but it is directly involved in their development. This approach is based on the inclusion of all interests and on the recognition of the need to move from a competitive process of negotiation of interests to a collaborative process of constructing consent. According to some (Bell et al., 2005; Wolsink, 2007; Longstaff et al., 2015), this approach would avoid situations in which a minority of people who do not support the proposed developments can paralyse the decision-making process.

Deliberative approaches to renewables include citizens' juries, interactive panels, workshops and conferences. However, the use of these tools in reality is extremely rare. Two examples can be found in the UK. The first concerns the development of the 2003 UK Government Energy White Paper, which provided for the adoption of all three of the public involvement strategy levels examined above, from simple information to complex decision-making processes (Haggett, 2011, p. 18). A similar procedure was also adopted in 2009, when residents of nine communities in England, Wales and Northern Ireland were invited to participate in a citizens' forum to evaluate a variety of options for insulating, heating and providing electricity to individual and community residences (Ipsos MORI, 2009, cited in Haggett, 2011, p. 18).

Although deliberative forms of public involvement are desirable, the literature on this issue also highlights some problematic aspects. The critical issues of deliberative and cooperative processes have to do with the power differentials that can emerge between social groups as well as with the issue of representativeness, that is, if and to what extent all the interests involved can be equally represented (Haggett, 2011). Indeed, even if collaborative planning methods are used, people will have access to different forms of influence and co-optation, based on factors such as language, education, social position, ethnicity and gender.

3.5 From Backyard to Territory: The Issue of Territorialisation of Renewable Energies

Initially, sociological research on res focused on the characteristics of the technologies involved in an attempt to identify which real or symbolic characteristics generated local opposition. This approach was dominant in many applied studies conducted during the 1990s which had as their object the public perception of the various energy sources (Devine-Wright, 2011, p. 57).

More recently, as we have seen in the previous paragraphs, the research has shifted its focus to the process of implementing technologies, on the basis that the local opposition is not caused as much by the characteristics of the technology itself as by top-down procedures that do not guarantee significant opportunities for local residents to participate in or benefit from projects.

A growing portion of the literature underlines the importance of developing a third research approach. The idea is that both lines of research have tended to marginalise a third crucial aspect regarding where these projects are developed. It seems increasingly important to seriously consider the *spatial dimension* of the energy transition (Bridge et al., 2013; Neal, 2013; Osti, 2013).

Two dominant approaches in the analysis of the "where" of the energy transition can be identified. The first, already examined above, considers the places of renewable energy production essentially as "backyards" dominated by NIMBY oppositions. The second regards the projects on res as "sites to be developed". This is the perspective mainly chosen by the experts. The technical language focuses on the identification of salient objective characteristics of potential locations, like the average annual wind speed, proximity to the electricity grid, accessibility, land ownership, and so on (Devine-Wright, 2011, p. 59).

Both the perspective of the NIMBY approach discussed above and that centred on the location of the sites appear to be problematic. As supported by a series of authors (Devine-Wright, 2011; Bridge et al., 2013) a preferable representation is that which considers the "where" of the energy transition through the strongly overlapping concepts of place, territory and landscape, understood as assemblages of cultural, social and geographical characteristics related to the production of energy and its consumption through space.

The representations centred on the backyard or on the site reproduce a vision of the place as impoverished and do not take into consideration the multiple meanings of the territory for individuals and communities, as well as the characteristics of the relational structure. The locations for res projects are not only sites with topographical, ecological or archaeological features but also places interwoven with symbolic or emotional elements (memories, stories and myths), as well as relationships and social capital.

A number of empirical studies (e.g. Van der Horst, 2007; Devine-Wright, 2009; Devine-Wright & Howes, 2010) claim that public opposition occurs more frequently in social contexts in which attachment to the place and identification with it are altered. This is also linked to the fact that the meanings associated with a certain place allow a sense of continuity and of individual effectiveness to be built over time. To the extent that local actors consider a project to have a negative impact on relational continuity, identity or individual/collective efficacy, they can adopt oppositional actions aimed at protecting the sense of place.

Osti (2013) too considers the landscape—understood as a combination of space, nature and culture—as a possible key for interpreting the ambivalence of renewable sources and a conceptual tool for promoting its sustainability. According to the author, when renewable energy facilities are integrated into a local image of space and landscape, it is more likely that they are accepted. On the contrary, when the local community, particularly in fragile rural areas, sees the original image of its territory fading, even res take the form of a threat to local development and predatory colonisation by external forces.

3.6 Conflicts Concerning Renewable Energies in Italy

Italy is not new to conflicts over res. The combination of a generous incentive system for renewable energy and the liberalisation of the electricity market in the late 1990s produced new tensions related to land use.

Data collected by various observatories on local conflicts such as the NIMBY Forum (https://www.nimbyforum.it/) or the Italian Atlas of Environmental Conflicts (http://cdca.it/atlante-italiano-dei-conflitti/) show that in the past ten years mobilisations related to energy have become the most frequent type of local environmental controversy in Italy. Among them, although the controversies surrounding the extraction of hydrocarbons on land and sea have not seemed to diminish, conflicts over renewables have increasingly gained in importance.

The conflicts over renewable energy in Italy present a spatially differentiated pattern linked to the differentiated diffusion of production facilities in the areas of the country (Osti, 2013). In particular, central and southern regions of Italy have seen large private companies buy or rent fields to install medium/large-sized photovoltaic panels on the ground (greater than 1 MW) or large wind farms. As shown by Carrosio and Scotti on wind farms (2019), very often renewables facilities in these areas are the result of initiatives of multinational companies, which are almost totally disconnected from the local agriculture and local society. Their only goal is to supply the markets and receive subsidies. Accordingly, a lack of public involvement in decision processes on wind farms can generally be highlighted. The same can be said for photovoltaic parks.

Photovoltaic technology creates tensions related to land use and in particular to the competition between green energy production and agricultural production. For wind, tensions emerge especially when the wind farms are located in places where initiatives of on-farm tourism have been developed. In this situation, wind ends up competing with a different model of local development that seeks safer sources of income.

It has been highlighted, however, that the opposition to large photovoltaics on the ground and wind power appears very weak in the regions of the centre-south of Italy. A number of studies (Agustoni & Sanseviero,

2011; Maretti, 2011) stress that the development of large wind farms has occurred in contexts that present peculiar characteristics from the point of view of political anthropology, such as a political tradition permeated by the ambiguities of clientelism and a social fabric characterised by medium/low intensity social capital. The result is that in many situations the installation of wind farms seems to occur without causing particular local opposition from organised civil society. When there is opposition, it is rarely endogenous but is fuelled almost exclusively by national environmental associations which organise their campaigns based mainly on the protection of biodiversity.

The situation in Northern Italy appears to be very different. In particular, the Po Valley, characterised by intensive agriculture and widespread urbanisation, has seen a widespread diffusion of large biogas plants. According to the data of the NIMBY Forum Observatory, most of the conflicts on res in Italy concern biomass, and especially biogas. Moreover, recently a widespread movement of opposition to mini-hydroelectric facilities has developed in the Alps. In the following, drawing on original empirical research by the authors, the conflicts that emerged around these two sources will be analysed in depth with particular attention to determining how they challenge NIMBY definitions.

3.6.1 Conflicts Over Biomass

The term biomass defines those raw materials that are obtained both from the waste of agricultural, zootechnical and forestry activities and from crops expressly dedicated to the production of energy.

Biogas plants are concentrated in the Northern Italy where there is wider access to farm residues and corn. The electricity produced is transferred to the network, which is not controlled locally, while the heat produced is generally dispersed or partially used by the company itself. Usually biogas plants do not contribute to strengthening the relationships with the local community, but on the contrary, in most cases, the population is hostile towards these structures especially when—as often happens—they are of a large size (more than 1 MW). Plants of a smaller size (250–500 kW), combined with the breeding activity, are less opposed by local communities as they represent a way to process agricultural waste, which is otherwise difficult to manage and impacts on the environment and tourism (Osti, 2013).

The first biogas plants in Italy were designed as systems to collect and treat potentially polluting waste generated by agricultural activity (Carrosio, 2013). However, the incentives have pushed towards a rethinking of biogas as a form of production of goods in itself. This organisational evolution has fuelled, since the early 2000s, a strong opposition movement in Northern Italy.

In 2012, starting from some committees of Emilia-Romagna, a national committee was set up against biogas plants, which collected the membership of committees from ten regions. The controversies surrounding biogas present some recurring narratives that are interesting to highlight. As argued by Magnani (2012b), those who promote the project to local communities generally adopt a weak ecological modernisation discourse, which presents the plant as the only solution available to solve the environmental problem—locally and globally—linked to agriculture and intensive farming and as the only system capable of sustaining itself economically through the business of renewable energies.

In particular, this discussion is articulated in some frequently used storylines. First of all, the promoters underline the plant's ability to solve the nitrogen pollution problem of land and water. It is highlighted how anaerobic digestion, allowing the improvement of the agronomic characteristics of the slurry and manure, ultimately allows the environmental impact of agricultural activity to be reduced (Verdi et al., 2019). A second storyline concerns the biogas plant's ability to affect the global environmental problem of climate change and to contribute to the achievement of greenhouse gas reduction. It is argued that, unlike traditional open-air manure storage techniques, the biogas plant allows the release of methane gas into the atmosphere to be avoided. At the same time, it is capable of producing significant energy savings.

A third discourse focuses on the economic benefits for local agriculture deriving from the plant. In particular, the self-production of electricity is shown as an important opportunity for the integration of agricultural income.

Opposition committees focus on showing the limits of this weak vision of the ecological modernisation of local agriculture. This is done not only by raising health issues related to the production of dioxins and other emissions deriving from combustion processes for the production of electricity but also by focusing on the issue of raw materials used to produce biogas and their energy efficiency. It is emphasised that the waste from agricultural activity, in particular sewage, is not very energy efficient. The consequence is that farmers who want to manage large plants are inclined to intensify their business, increasing farm size as well as growing corn for energy. The demand for maize and other cereals pushes farmers towards monoculture and thwarts attempts to encourage diversification and multifunctionality (Osti, 2013).

Another type of biomass of particular relevance for its diffusion is wood, used for heating through district heating networks (Carrosio & Scotti, 2019; Carrosio & Magnani, 2020). As argued by Osti (2013), compared to biogas, this use of resources for the production of renewable energy seems to be an example of success. The incentives for the exploitation of forest biomass have in fact created new job opportunities in the forestry sector and in the maintenance of district heating systems in rural mountain areas, often characterised by high unemployment and depopulation rates. However, new tensions have emerged especially with environmental associations regarding the issue of biodiversity conservation, identified by semi-abandoned forests on the one hand and urban air pollution resulting from an increasing use of wood for domestic heating on the other.

In this case as well, a fundamental problem is represented by the dimensions of the facilities. As pointed out by Osti (2013), large-scale district heating plants have a wide environmental footprint and require a quantity of biomass (pellets or wood) not supplied on site and which can compete with food and agriculture in other regions of the country or the world. Furthermore, large wood biomass plants for the production of thermal and electrical energy may also come into conflict with the diffusion of other res, especially solar thermal and photovoltaic.

Moreover, as highlighted by Carrosio and Magnani (2020), the dimension of the facilities is often closely related to the distinction between heat networks in rural areas and in urban centres. In rural areas, district heating networks are mostly managed by companies highly embedded with local actors. They are fed by biomass supplied by local forest operators and sawmills. Also, the users participate as shareholders of the company that provide heat. In urban areas, the networks are instead managed by multi-utilities, and they are fed by a variety of resources, including fossil ones. The users in this case are passive consumers.

Particularly interesting in this context seems to be the case of South Tyrol—an Italian Autonomous Province—where the presence of institutionalised and community networks seems to ward off some of the negative aspects connected to the biomasses highlighted here.

As pointed out by Puttilli (2014, p. 145), South Tyrol has been able to exploit the forest resources through territorial solutions capable of extending the benefits of this energy resource beyond the energy field. This has been the result of a series of favourable social conditions. First of all, one has to consider the role of the Autonomous Province, which since 1993 has adopted a policy of strong incentives for biomass plants. Secondly, the organisational capacity of the actors of the forests and energy sectors, capable of building an efficient and deep-rooted network at the valley level between forestry companies, sawmills and energy plants, has to be highlighted. Thirdly, the tradition that characterises this area in forming consortia or cooperative enterprises that bring together municipalities, companies and citizens in the management of biomass plants has to be stressed.

3.6.2 Conflicts Over Hydroelectric Energy

Increasing pressures also concern water resources. In the first half of the last century, most of the Italian waterways were used to produce hydroelectric energy with the construction of large-medium dams. Since 2008, as an effect of the high remuneration offered by incentives for renewables (three times the market value for a period of 20 years), the exploitation of residual water courses has resumed (Legambiente, 2017).

At the local level, in recent years, committees aimed at opposing mini-hydroelectric projects in small rivers have multiplied, especially in the mountain areas of the Alps and the Apennines. In these cases, as often happens in environmental conflicts, we find the formation of transversal alliances that unite diversified interests around the river. In particular in these controversies, environmental associations are found side by side with those actors, such as fishing associations and rafting centres, for whom the river represents a place to carry out recreational activities, as well as with citizens who share a strong sense of belonging to a common territory and a local landscape.

In 2017, the various committees and associations that had fought for the defence and protection of rivers reunited in the "National Coordination for the Protection of Rivers—Free Rivers Italy". The objective of the coordination is to exert pressure on the Italian government and the regions to safeguard the rivers of the Alps and the Apennines which are still free from hydroelectric and irrigation.

As highlighted by Magnani (2021) anti-mini-hydro movements challenge the NIMBY definition by promoting a peculiar form of environmentalism. In particular, at the discursive level, such mobilisations, far from being just focused on the localisation of the plants, articulate a broader discourse opposing mini-hydro as climate change mitigation strategy and promoting a vision of the river as biodiversity to be protected rather than as energy to be exploited. At the heart of the committees' claims, there is the argument that if it is true that the new hydroelectric plants do not release polluting agents into the atmosphere or water, it is equally true that they deplete or preclude the vital functions of the waterways, already under pressure from the effects of climate change.

Furthermore, it is stressed that these environmental costs cannot be justified by the benefits deriving from the energy supply that could be produced. In fact, it is argued that even if all the approximately 2000 plant projects under investigation in Italy were carried out, the percentage of energy produced would be extremely limited, equal to 2‰ (2 per thousand) of the total energy consumed in Italy in one year. The committees, therefore, call for water courses, and in particular mountain ones, to be considered the heritage of biodiversity, environmental and landscape values to be protected rather than exploited intensively. However, this does not imply denying that rivers are a precious resource for the landscape, capable of promoting alternative and less impactful recreational tourism, allowing a harmonious economic and social development of the territory.

These objectives are reflected in actions aimed at ensuring that the management plans of the river basin districts establish savings policies in the use of the water asset and at the same time envisage measures aimed at the requalification of watercourses. In particular, in the management of local water resources, the need to overcome the current concept of minimum flow, used to decide whether or not to authorise a specific plant, in favour of advocating the concept of ecological flow, introduced by European legislation. The latter aims not only to guarantee the maintenance of the ecological quality objectives of a watercourse but also to safeguard the ecosystem services performed by it for the territories in which it is located.

Eventually, independent of the outcome of individual controversies, environmental conflicts over mini-hydroelectric facilities are particularly interesting from a sociological point of view because they seem to activate reterritorialisation processes by local communities as well as new processes of socialisation concerning rivers in mountain areas (see Magnani, 2021).

3.7 Conclusion

This chapter has dealt with a particular kind of civil society participation in the energy transition that sees the public as a social actor in conflicts concerning the construction and location of renewables production facilities. In relation to that, in the first part of the chapter, we started from a critical deconstruction of the NIMBY syndrome and its tenets as an analytical construct to understand environmental conflicts related to renewables.

In particular, we have stressed that a number of empirical studies have shown the groundlessness of the assumptions about the public implicit in the term. Accordingly, we have then highlighted the relevance of other analytical constructs, like the concept of environmental justice, the related issues of participation and consensus and the dynamics of territorialisation to understand local opposition to renewables.

In the second part of the chapter, we have considered the most important conflicts over renewable energy sources in Italy, and we have highlighted the key social dynamics that characterise them.

We have seen that despite wind and photovoltaics having mostly been developed in a *deterritorialised* fashion in the centre-south of the country, no important conflict has emerged concerning these sources.

The situation is different with regard to biomasses and hydropower in the north of the country.

Here our analysis has shown that the conflicts considered challenge NIMBY definitions in different ways. In particular, with regard to biogas, the oppositional movement that has developed did not oppose specific localisations of the facilities but on the contrary elaborated a wider critique of the technology proposed and the underlying discourse of ecological modernisation of agriculture.

Moreover, concerning mini-hydroelectric facilities, citizens did not oppose specific localisations but rather the dominant representation of the mini-hydroelectric technology as a climate change mitigation solution.

Issues of environmental justice concerning equal distribution of costs and benefits were also relevant in the conflicts considered. In both cases, the interests of a few proponents taking advantage of incentives for energy production were seen as counterposed to the interests of the many people inhabiting the places and landscapes. The cases analysed were also lacking from the point of view of procedural justice. Neither in conflicts over biogas plants nor in conflicts over mini-hydroelectric power were significant forms of public involvement in decision-making over facilities were adopted.

Eventually in both cases, the oppositional alliances that emerged were characterised by a strong attachment to the territory. This highlights the need for a planning perspective which no longer aims to ensure the acceptance of a technology by the population but to identify the ways in which the technology can be adapted to the place and the collective sense of belonging.

References

Agustoni A., & Sanseviero S. (2011). La forza del vento. Il conflitto sull'installazione di impianti eolici in Abruzzo. In L. Pellizzoni (a cura di), *Conflitti ambientali* (pp. 97–127). Il Mulino, Bologna: Esperti, politica, istituzioni nelle controversie ecologiche.

Aitken, M. (2010). Why we still don't understand the social aspects of wind power: A critique of key assumptions within the literature. *Energy Policy, 38*(4), 1834–1841.

Arnstein, S. R. (1969). A ladder of citizen participation. *Journal of the American Institute of Planners, 35*(4), 216–224.

Bell, D., Gray, T., & Haggett, C. (2005). The 'social gap' in wind farm siting decisions: Explanations and policy responses. *Environmental Politics, 14*(4), 460–477.

Bell, D., Gray, T., Haggett, C., & Swaffield, J. (2013). Re-visiting the 'social gap': Public opinion and relations of power in the local politics of wind energy. *Environmental Politics, 22*(1), 115–135.

Bellaby, P. (2010). Uncertainties and risks in transitions to sustainable energy, and the part 'trust' might play in managing them: A comparison with the current pension crisis. *Energy Policy, 38*(6), 2624–2630.

Breukers, S., & Wolsink, M. (2007). Wind power implementation in changing institutional landscapes: An international comparison. *Energy Policy, 35*(5), 2737–2750.

Bridge, G., Bouzarovski, S., Bradshaw, M., & Eyre, N. (2013). Geographies of energy transition: Space, place and the low-carbon economy. *Energy Policy, 53*, 331–340.

Burningham, K. (2000). Using the language of NIMBY: A topic for research, not an activity for researchers. *Local Environment, 5*(1), 55–67.

Burningham, K., Barnett, J., & Walker, G. (2015). An array of deficits: Unpacking NIMBY discourses in wind energy developers' conceptualizations of their local opponents. *Society & Natural Resources, 28*(3), 246–260.

Carrosio, G. (2013). Energy production from biogas in the Italian countryside: Policies and organizational models. *Energy Policy, 63*, 3–9.

Carrosio, G., & Magnani, N. (2020). District heating and ambivalent energy transition paths in urban and rural contexts. *Journal of Environmental Policy & Planning, 22*, 1–13.

Carrosio, G., & Scotti, I. (2019). The 'patchy' spread of renewables: A socio-territorial perspective on the energy transition process. *Energy Policy, 129*, 684–692.

Chilvers, J., Pallett, H., & Hargreaves, T. (2018). Ecologies of participation in socio-technical change: The case of energy system transitions. *Energy Research & Social Science, 42*, 199–210.

Devine-Wright, P. (2005). Beyond NIMBYism: Towards an integrated framework for understanding public perceptions of wind energy. *Wind Energy, 8*(2), 125–139.

Devine-Wright, P. (2009). Rethinking NIMBYism: The role of place attachment and place identity in explaining place-protective action. *Journal of Community & Applied Social Psychology, 19*(6), 426–441.

Devine-Wright, P. (2011). From backyards to places: Public engagement and the emplacement of renewable energy technologies. In P. Devine-Wright (a cura di), *Renewable energy and the public: From NIMBY to participation* (pp. 57–70). Earthscan.

Devine-Wright, P., & Howes, Y. (2010). Disruption to place attachment and the protection of restorative environments: A wind energy case study. *Journal of Environmental Psychology, 30*(3), 271–280.

Devine-Wright, P., & Wiersma, B. (2013). Opening up the "local" to analysis: Exploring the spatiality of UK urban decentralised energy initiatives. *Local Environment, 18*(10), 1099–1116.

Firestone, J., Kempton, W., & Krueger, A. (2009). Public acceptance of offshore wind power projects in the USA. *Wind Energy, 12*(2), 183–202.

Freudenburg, W. R., & Pastor, S. K. (1992). NIMBYs and LULUs: Stalking the syndromes. *Journal of Social Issues, 48*(4), 39–61.

Funtowicz, S. O., & Ravetz, J. R. (1993). Science for the post-normal age. *Futures, 25*(7), 739–755.

Gross, C. (2007). Community perspectives of wind energy in Australia: The application of a justice and community fairness framework to increase social acceptance. *Energy Policy, 35*(5), 2727–2736.

Gross, M., & Mautz, R. (2015). *Renewable energies*. Routledge.

Haggett C. (2011). 'Planning and persuasion': Public engagement in renewable energy decision-making. In P. Devine-Wright (a cura di), *Renewable energy and the public: From NIMBY to participation* (pp. 15–27). Earthscan.

Jenkins, K., McCauley, D., Heffron, R., Stephan, H., & Rehner, R. (2016). Energy justice: A conceptual review. *Energy Research & Social Science, 11*, 174–182.

Jobert, A., Laborgne, P., & Mimler, S. (2007). Local acceptance of wind energy: Factors of success identified in French and German case studies. *Energy Policy, 35*(5), 2751–2760.

Kasperson, R. E., & Ram, B. J. (2013). The public acceptance of new energy technologies. *Daedalus, 142*(1), 90–96.

Legambiente. (2017). L'idroelettrico, impatti e nuove sfide al tempo dei cambiamenti climatici. https://www.legambiente.it/contenuti/dossier/

Longstaff, H., Secko, D. M., Capurro, G., Hanney, P., & McIntyre, T. (2015). Fostering citizen deliberations on the social acceptability of renewable fuels policy: The case of advanced lignocellulosic biofuels in Canada. *Biomass and Bioenergy, 74*, 103–112.

Luloff, A. E., Albrecht, S. L., & Bourke, L. (1998). NIMBY and the hazardous and toxic waste siting dilemma: The need for concept clarification. *Society & Natural Resources, 11*(1), 81–89.

Magnani, N. (2012a). Exploring the local sustainability of a green economy in Alpine communities. A case study of conflict over a biogas plant. *Mountain Research and Development, 32*(2), 109–116.

Magnani, N. (2012b). The green energy transition: Sustainable development or ecological modernization? *Sociologica, 2*, 1–25.

Magnani, N. (2021). Beyond NIMBY: Mobilization against mini-hydroelectric power in the Italian Alps. *Mountain Research and Development, 40*(3), 31–36.

Maretti, M. (2011). Governance e desiderabilità sociale delle energie alternative nel caso abruzzese. In L. Pellizzoni (a cura di), *Conflitti ambientali* (pp. 129–145). Esperti, politica, istituzioni nelle controversie ecologiche.

Nadaï, A. (2007). "Planning", "siting" and the local acceptance of wind power: Some lessons from the French case. *Energy Policy, 35*(5), 2715–2726.

Neal, S. (2013). Transition culture: Politics, localities and ruralities. *Journal of Rural Studies, 32*, 60–69.

Osti, G. (2013). Land use tensions for the development of renewable sources of energy. In S. Lockie, D.A. Sonnenfeld, & D.R. Fisher (a cura di), *Routledge international handbook of social and environmental change* (pp. 319–330). Routledge.

Pallett, H., Chilvers, J., & Hargreaves, T. (2019). Mapping participation: A systematic analysis of diverse public participation in the UK energy system. *Environment and Planning E: Nature and Space, 2*(3), 590–616.

Pasqualetti, M. J. (2011). Opposing wind energy landscapes: A search for common cause. *Annals of the Association of American Geographers, 101*(4), 907–917.

Pellizzoni, L. (a cura di) (2011). *Conflitti ambientali.* Esperti, politica, istituzioni nelle controversie ecologiche.

Puttilli, M. (2014). *Geografia delle fonti rinnovabili.* Energia e territorio per un'eco-ristrutturazione della società.

Rayner, S. (2010). Trust and the transformation of energy systems. *Energy Policy, 38*(6), 2617–2623.

Upreti, B. R. (2004). Conflict over biomass energy development in the United Kingdom: Some observations and lessons from England and Wales. *Energy Policy, 32*(6), 785–800.

Van der Horst, D. (2007). NIMBY or not? Exploring the relevance of location and the politics of voiced opinions in renewable energy siting controversies. *Energy Policy, 35*(5), 2705–2714.

Verdi, L., Kuikman, P. J., Orlandini, S., Mancini, M., Napoli, M., & Dalla, M. A. (2019). Does the use of digestate to replace mineral fertilizers have less emissions of N2O and NH3? *Agricultural and Forest Meteorology, 269–270*, 112–118.

Walker, G., & Devine-Wright, P. (2008). Community renewable energy: What should it mean? *Energy Policy, 36*(2), 497–500.

Wolsink, M. (2006). Invalid theory impedes our understanding: A critique on the persistence of the language of NIMBY. *Transactions of the Institute of British Geographers, 31*(1), 85–91.

Wolsink, M. (2007). Planning of renewables schemes: Deliberative and fair decision-making on landscape issues instead of reproachful accusations of non-cooperation. *Energy Policy, 35*(5), 2692–2704.

Wüstenhagen, R., Wolsink, M., & Bürer, M. J. (2007). Social acceptance of renewable energy innovation: An introduction to the concept. *Energy Policy, 35*(5), 2683–2691.

Wynne, B. (1996). May the sheep safely graze. A reflexive view of the expert-lay knowledge divide. In S. Lash, B. Szerszynski, & B. Wynne (a cura di), *Risk, environment and modernity: Towards a new ecology* (pp. 44–83). Sage.

4

The Social and Spatial (In)justice of the Energy Transition Policies

Giovanni Carrosio

Abstract The energy transition takes on various connotations according to the territorial contexts and unequally involves different social groups. In particular, energy transition policies favour or penalise specific places and people. The literature has addressed the issue of inequalities in two ways: first, by investigating conflicts for the location of renewable energy plants; second, in relation to the energy poverty issue, by highlighting how problems of access to energy as a fundamental good still exist. In this chapter, instead, we address the issue of inequalities in a third way: we want to investigate how the policies for the diffusion of renewable energies, for energy saving and energy efficiency, favour or disadvantage weaker social groups and marginal territories, producing or reproducing social and spatial inequalities.

Keywords Energy justice • Social inequalities • Spatial inequalities • Energy policies • Energy transition

G. Carrosio (✉)
Department of Social and Political Sciences, University of Trieste, Trieste, Italy
e-mail: gcarrosio@units.it

N. Magnani, G. Carrosio, *Understanding the Energy Transition*,
https://doi.org/10.1007/978-3-030-83481-4_4

4.1 The Question of Justice in the Energy Transition

Starting in the 1970s, environmental movements began to promote the transition from the production of energy from fossil sources to the production of energy from renewable sources, thinking that the energy transition, as well as being desirable for environmental reasons, could also promote a political transition towards more radical democracies. It is what in literature today is called "just transition" or "just sustainability". Environmentalists and researchers working in the field of renewable energies thought that technologies for transforming renewable sources into energy gave rise to highly decentralised energy systems, which citizens and local communities would appropriate. The idea of an imminent democratisation of the economic system and a growing empowerment of the territories, according to the principles of municipalism and solidarity federalism (Magnaghi, 2020), was widespread. This politicisation of the energy transition was very strong during the collective movements against nuclear power. The nuclear system, thus centralised in large plants and linked to expert knowledge that was not intelligible to citizens, was opposed to renewable energy as simple and naturally distributed technologies. This way of looking at the energy transition faded over time. In particular, in Italy, nuclear power was abandoned following a popular referendum in 1987. It was the last real battle carried out by the environmental movement, which, in the following years, began its path of institutionalisation. Italian environmentalism assumed a mode of corporate action, entering the governance of the environment and dismissing the repertoire of action of the protest. By becoming part of the establishment, most of the environmental movement did not invest much energy in combining the energy transition with a political transformation towards the democratisation of energy. Starting in the 1990s, environmental organisations also began to pursue the energy transition within the paradigm of *ecological modernisation*. Combining its action with the growing emergence of the problem of climate change, the main problem became how to accelerate the transition, regardless of the nature of the economic actors who interpreted it and the political-organisational forms with which renewable energies would spread. The paradigm of ecological modernisation (Spaargaren, 2000), which informs European policies on the environment, intends to promote the energy transition within the traditional development model. It has great faith in capitalism and technological innovation, putting large urban agglomerations at the centre of the transition. It wants to solve environmental problems through strategies based on economic growth (Matlock & Lipsman, 2020), such as tax policies to encourage the market demand for ecological products, tax exemptions for energy saving interventions, incentives for the production of energy from renewable sources and real estate policies for the enhancement of high-energy class homes. This modernisation ecological paradigm, arguing that efficient technologies and regulated growth can produce a "green capitalism", does not take into consideration the social and territorial effects of the energy transition (Carrosio, 2020a). It is for these reasons that today it is evident that a "just transition" is not something that emerges automatically from the replacement of fossil fuels with renewable energies. As some scholars point out, those of ecological modernisation are socially and territorially blind policies (Barca et al., 2012), which pursue the goal of decarbonising production systems, without worrying about the social and territorial consequences. Consequently, by not taking into account the social and territorial diversities on which they act, they contribute to reproducing social and territorial inequalities. For this reason, we can no longer think that social and territorial justice can be technologically determined because the changes that low-carbon transition may bring could be more exploitative and socially detrimental than pre-transition systems (O'Sullivan et al., 2020). A contemporary strand of the energy transition literature investigates precisely these aspects, looking at the consequences of energy policies with respect to different types of justice (Sovacool & Dworkin, 2015). However, while distributional, recognition and procedural justice are well represented in the literature (Jenkins et al., 2016), little attention is paid to the spatialisation of energy justice (Bouzarovski & Simcock, 2017). In this chapter, we will focus in particular on the distributive and territorial injustices deriving from energy transition policies. In the first paragraph, we will address the social distortions of policies, showing how they reproduce and increase socio-economic inequalities. In the second paragraph, however, we will address the spatial distortions of policies, highlighting territorial inequalities. In doing this, we will draw on examples relating to the energy transition in Italy. Finally, we will try to suggest how energy transition policies could be bent to combat social and territorial inequalities.

4.2 Social Inequalities in the Italian Energy Transition

Starting from the early years of the twenty-first century, also under the influence of European directives on energy, Italy has begun to promote the energy transition through a series of very important public policies. In particular, the policymaking activity focused on the promotion of new plants for the production of energy from renewable sources and on the energy efficiency of homes and industrial processes (Carrosio, 2015). These policies, together with many other types of intervention, by acting on the supply and demand of energy, have undoubtedly triggered the energy transition process. This is demonstrated by the fact that by 2020, 40% of the demand for electricity in Italy was covered by renewable sources and that overall the Italian energy system (heat, electricity, and transport) has reached about 20% coverage with renewable energy. However, Italy, despite being in line with the European objectives, remains far from the ambitious goal of total decarbonisation. Furthermore, the transition is not producing a substantial modification of the development model but is completely within the paradigm of ecological modernisation.

This paradigm believes in the compatibility between the growth in the volume of traded goods and sustainability. This is possible thanks to the role of technological innovation, which should make production systems more efficient. Most of the instruments adopted by the public to stimulate the transition are confident about this paradigm: they are based on the model of market-based policies, which aim to achieve decarbonisation objectives through regulation and incentive systems (Stavins, 2003).

The combination of technological innovation and the market has given rise to policies which, for example, have encouraged the production of energy from renewable sources and the adoption of devices for the energy retrofit of the building stock, have leveraged the building certification mechanism to stimulate energy improvement and encouraged the purchase of electric or hybrid cars. As they were built, these policies were differently accessible by citizens. The choice was made to boost the energy transition by stimulating innovation in the middle-upper class: homeowners, families with stable incomes, savings and spending capacity to face major investments, such as the installation of photovoltaic panels and structural interventions for energy saving in one's home, the purchase of a new car with ecological characteristics, the purchase of homes in high-energy classes. For various reasons, the lower and middle classes have had difficulty accessing these policies. Incapable, precarious, low-income families with no savings and tenants are the types of situations in which people live, who for different reasons have problems accessing policies based on direct incentives or in the form of tax deductions (Carrosio, 2020b). At the same time, it is those social classes that more than others express skepticism about the environmental issue, as a consequence of the fact that they do not see the tangible link between the transition and an improvement in the quality of their life (Gottenhuber & Mulholland, 2020).

In particular, two initiatives that have had an important influence on the structuring of the political-organisational field of the energy transition in Italy should be noted. They help us to unveil the relationship between energy transition and social and territorial inequalities.

The first concerns incentives for the production of electricity from photovoltaic sources. The incentive system is based on remuneration for the energy produced through a tariff that has made this renewable source very attractive for financial investment by large groups and wealth families. The second concerns the system of tax deductions to encourage energy retrofitting of homes. Through a series of readjustments that have changed deduction thresholds over time, the policy instrument provides the opportunity to deduct expenses incurred to reduce the energy needs of homes.

These two seemingly socially and spatially neutral ways of promoting transition have had very different impacts. From a social point of view, they have concentrated the benefits in the wealthiest social classes: families with economic capital sufficient to make investments in energy production plants or in efficiency measures. From a territorial point of view, they have produced a search for spaces to build photovoltaic systems on the ground. These spaces have been found in the weakest territories, rural areas and marginal territories. In these areas, external entrepreneurs have made large investments disconnected from the local development of the territories. In the next paragraphs, we will use the case of the energy retrofit of homes to shed light on social disparities. The case of the spread of photovoltaics, on the other hand, will be used to discuss the issue of territorial disparities.

4.3 Social Injustice: The Case of Policies for the Energy Efficiency of Homes

Since 2008, one of the most important policy devices to promote savings and energy efficiency of homes was that of tax deductions on retrofit costs incurred by household. In particular, families were able to deduct 65% of their expenditure from their taxable income. This sum was recovered in ten years: every year, for ten years, the state recognised a tax deduction for the taxpayer.

Families could do various types of interventions: thermal insulation of the house, replacement of windows and the boiler, installation of solar heating systems, construction of underfloor heating systems, and so on. The interventions had to be certified and approved by National Agency for New Technologies on Energy (ENEA), a national public agency that deals with innovation in the energy sector. Starting from 2020, the Italian government has modified this policy, approving a decree that provides for the possibility of deducting 110% of the expenditure incurred, or to transfer the tax credit generated by the retrofit intervention to third parties. On this new policy, however, we do not yet have data available. Of the 65%, however, we can undertake an analysis based on a robust set of data, ranging from 2008 to 2017.

According to many observers, the 65% mechanism had been a successful tool to counter the crisis in the construction industry (Carrosio, 2015). In fact, tax deductions also had the objective of acting on the contraction of the construction market, making economically advantageous those interventions that otherwise would not be made following the current rules of the market. According to a Social and Economic Research Center (CRESME) study, between 2008 and 2017—the years in which the incentive tool was activated—tax deductions transported 34 billion euro of investments, allowing the loss of employment in the construction sector to be curbed. In ten years, more than 12 million energy retrofit interventions were carried out, for a total expense of approximately €11 billion in tax deductions. According to some estimates (CRESME, 2017), the incentive economy has allowed about 100,000 workers each year not to lose their jobs. The tax deductions sustained the construction sector but did not produce structured global retrofits of buildings, essentially for three reasons: they were used as a means of modernising homes, especially with partial interventions to replace the fixtures and boilers for heating; they have incentivised individual houses rather than condominiums; there is a growing segment of the population that, due to tax incapacity, does not find the deduction system attractive and therefore does not want to use any kind of incentive to make interventions in their own homes. This last factor requires a deepening. In fact, one of the risks of policy tools that incentivise people's behaviour is that they act unequally on the population. In this case, they redistribute wealth from the less wealthy to the richer classes. Let's look at the graph on the deductions of the last ten years (Fig. 4.1).

As shown by the graph, in percentage terms, the higher income classes have benefited more than the lower income classes from the deductions, both in terms of the number of deductions and the amount of money deducted. Both the increase of the quota deducted and that of the frequency of deductions are directly proportional to the growth of the income classes. This happens both on account of a greater propensity for and possibility of making investments and on account of a greater ability to access useful knowledge to obtain deductions. Inequality in access possibilities must be integrated with the theme of capabilities. There is not only the problem of household purchasing capacity, but there is also the problem of a lack of cognitive tools. These empirical evidences require

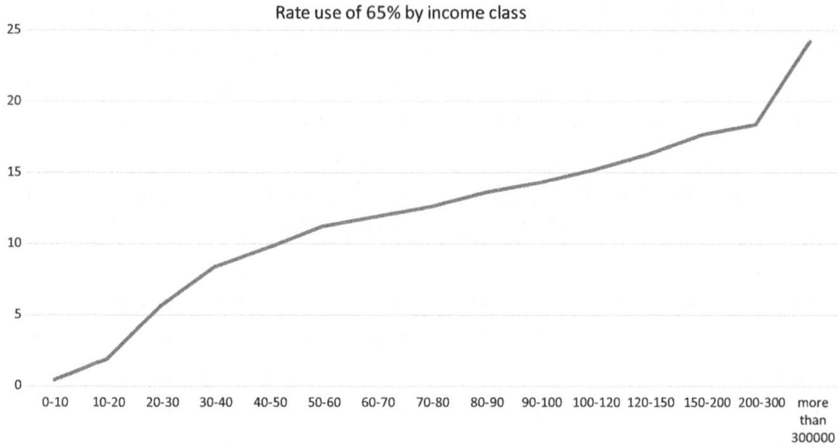

Fig. 4.1 Rate use of 65% by income class. The classes are in thousand euros. (Source: Our processing of Ministry of Finance and Economy data)

reflection on the relationship between environmental policies and inequalities, an issue often dealt with as regards *eco-gentrification* but which in this case also emerges from the use of market tools to change people's behaviour.

There are not, however, only social inequalities. If we compare income investments on a regional scale, we see important territorial differences. Obviously, the value of investments must first be standardised for the heating degree day (HDD)—an indicator designed to quantify the demand for energy needed to heat a building—and for the total net available income (NAI). In this way, we avoid the error of being influenced by climatic and wealth conditions in the reading of the spatial distribution of retrofit interventions.

With the same heating needs and disposable income, the distribution of the indicator indicates that the regions with the best performances are Veneto, Trentino Alto Adige and Valle d'Aosta. In general, all regions of the North have good performances, except Liguria and Piedmont. In the south, Sicily and Puglia stand out for better performances than other regions (Table 4.1).

Table 4.1 Investments on a regional scale standardised for the heating degree day and for the net available income

Region	HDD/NAI
Piemonte	137.7
Valle d'Aosta	211.4
Liguria	68.1
Lombardia	176.7
Trentino Alto Adige	212.3
Veneto	217.6
Friuli Venezia Giulia	197.4
Emilia Romagna	181.6
Toscana	145.6
Umbria	92.8
Marche	97.2
Lazio	70.1
Abruzzo	99.3
Molise	36.2
Campania	49.6
Puglia	74.4
Basilicata	111
Calabria	56.2
Sicilia	135.6
Sardegna	139.3

Source: Own calculations based on Ministry of Economy and Finance data, 2008–2017

We can define these differences as socio-territorial: factors such as the structure of companies, the capacity for innovation, social capital and local industrial paths are important.

On the other hand, upon analysing the types of intervention, we see how the tax deduction instrument was used above all for the modernisation of windows (Table 4.2).

The incentive system has rewarded the simplest and unitarily cheaper interventions compared to global interventions that are more expensive but more effective in terms of reducing consumption and emissions and with a more virtuous relationship between investment and energy savings. The replacement of the windows is in fact the least effective type of intervention from the point of view of energy saving and reduction of climate-altering emissions, and therefore more expensive in terms of the ratio of costs and savings achieved.

Table 4.2 Types of intervention with the tax deduction tool

Vertical walls	11.40%
Horizontal walls	18.30%
Windows	46.00%
Solar thermal	2.40%
Sunscreens	2.60%
Condensing boilers	14.90%
Geothermal systems	0.10%
Heat pump	3.10%
Water heaters	0.60%
Building automation	0.10%
Other	0.40%

Source: Own elaborations on ENEA, data 2008–2017

4.4 Territorial Injustice: The Case of the Spread of Photovoltaics in the Italian Provinces

Just as the policies for energy retrofit have reproduced and exacerbated the economic inequalities already existing in Italy, in the same way the spread of energy production from renewable sources has reproduced and exacerbated territorial disparities. In particular, if we consider photovoltaics, we see how the diffusion logics differ from place to place.

This dynamic has been analysed in other countries. For example, the work of O'Sullivan et al. (2020) highlighted that the decentralised development promised by the development of renewables in South Wales is not taking place in a way that emancipates local populations from marginalisation, but factors of economic inequality and policies make opportunities socially and spatially segregated. The transition becomes a further device for the peripheralisation of the most fragile rural areas. Places have differing abilities to engage with energy projects and innovations (Baker & Mehmood, 2015) so that pre-existing inequalities may be reinforced rather than rectified by the low-carbon transition. The ways of spreading photovoltaics in Italy tell us about this relationship between energy decentralisation and the incidence of structural factors of local economies and societies in the way of diffusion.

Table 4.3 Data on photovoltaic power plants in Italy up to December 2020

Kw	Number	Power MW	% power plants	% installed power MW
up to 3	297410	804	33.7931348	3.85352761
3–20	514162	3675	58.4215251	17.6140721
20–200	56302	4403	6.39730028	21.1033359
200–1000	11066	7504	1.25737141	35.9662577
more than 1000	1150	4478	0.13066845	21.4628067

Source: Our processing of Electricity services manager (GSE) data

Thanks to generous incentive policies for the production of energy from the sun, between the early 2000s and 2020 more than 800,000 photovoltaic systems were installed. Most of them are very small family plants (Table 4.3).

As shown in the table, plants up to 3 KW—certainly family ones—represent about 33% of the total but only an installed power of 804 MW (3.85% of the photovoltaic power installed in Italy). Plants between 3 and 20 kW, on the other hand, account for approximately 58% and represent 17% of the power. These plants are partly family owned and partly made up of small and medium-sized enterprises. Certainly industrial plants, partly funded by companies and partly by investment funds, are those above 200 KW. They represent approximately 1.3% of the total but 57.5% of the installed power. In the face of the extraordinary spread of small plants, which highlights the important activism of families and small businesses, most of the installed power is reached thanks to just over 1% of plants. We can therefore identify a double dynamic: a large diffusion of photovoltaic technology among small businesses and presumably upper-middle-class families and a large concentration of the wealth produced by photovoltaics in large companies and investment funds. In fact, the incentive system remunerates the energy produced through a levy in the bills of all Italian users in order to redistribute it to the production of energy on the part of the owners of the plants.

In addition to these differences in size and distribution, we can also identify geographical differences, which help us to thematise the relationship between energy transition and territorial inequalities. Let us start with the analysis of the data, which we were able to disaggregate at the provincial level Nomenclature of territorial units for statistics (NUTS) 3, following the EU classification of territorial levels). The provincial level is not the optimal setting for this type of analysis; however, it is detailed enough to reveal some territorial differences. The more than 800,000 photovoltaic systems are unevenly distributed in the provinces. The number and average size of the plants vary considerably. These data interest us because they bring out the level of distribution. If in a province we have on average small plants, this means that solar technology is spread more democratically among the citizens; if, on the other hand, we have large plants on average, we are faced with cases in which large companies have made important investments by centralising the control of the energy transition.

Considering the specific aspects of photovoltaic technology, one way to understand the degree of democratisation of energy production is given by the average size of the plants at the provincial level (Blasutig, 2017). If we divide the installed power by the number of plants, we can understand if small plants prevail in a territory or if there is a prevalence of large plants. In the first case, we have a high level of decentralisation with many producers; in the second case, we have a centralisation of production with the prevalence of a few large players. Secondly, if we correlate the average power with the per capita income, we can understand if the big players direct their investments to the weakest areas of the country. Certainly the relationship between GDP per capita and the weakness or strength of the provincial territories is not immediate, but it is still a good summary indicator, as shown by the works on the civil economy in Italy (Carrosio & Osti, 2011).

As can be seen from the graph (Fig. 4.2), there is an inverse correlation between per capita income and the average size of the plants (-0.5778). With a few minor exceptions, where the income is higher, the plants are smaller and therefore more widespread; where it is lower, the plants are larger and more concentrated. This evidence also assumes spatial connotations, which highlight the dynamics of territorial inequality fuelled by the way policies for the energy transition are constructed. The provinces where the dynamic takes on more marked connotations are those listed in Table 4.4.

Rovigo, Alessandria and Cuneo are among the provinces of Northern Italy with the highest rate of rurality. They have large spaces in which to

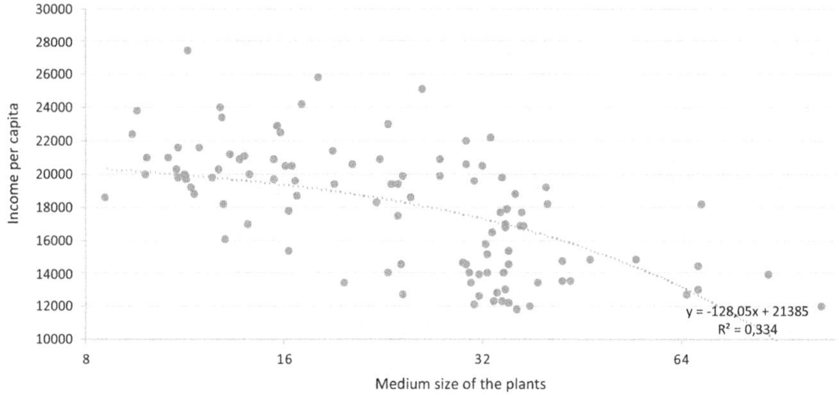

Fig. 4.2 Correlation between income per capita and medium size of the plants at the provincial level. (Source: Own elaboration of GSE 2020 data)

Table 4.4 Medium size of the solar plants in the provinces over 40 KW

Province	Medium size of the solar plants
Foggia	105.434307
Brindisi	87.2913977
Rovigo	69.269569
Barletta-Andria-Trani	68.4506319
Viterbo	68.225134
Matera	65.8194946
Taranto	54.9815087
Macerata	47.0394281
Campobasso	43.8680285
Lecce	42.5814024
Teramo	42.482244
Alessandria	40.4140216
Cuneo	40.129602

Source: Own elaboration of GSE 2020 data

place large photovoltaic systems, but they are also among the provinces with the lowest incomes of North Italy. The combination of low population density and the weakness of local institutions make these provinces places where external speculative investments are successful without opposition from the local civil society (see Chap. 3). In fact, in the province of Rovigo, the largest ground-mounted photovoltaic system in Italy (72 MW) was installed. It is a solar plant owned by an American investment fund, which used Spanish manpower for the construction and now produces energy without the need for local labour. It is managed remotely both for the technical aspects and for those concerning the site security. In Central-Southern Italy, Foggia, Brindisi and Barletta are three provinces with the most important presence of photovoltaics of the South. All of three are part of Puglia, the region with the highest installed power in Italy. Following the spread of large photovoltaic systems on the ground, some environmental movement campaigns against soil consumption and for the protection of the landscape were born throughout Italy. According to data from the NIMBY observatory, however, there is also a relationship between the mobilisation rate and territorial disparities. In fact, the citizens of the rich provinces are mobilised against these initiatives much more than the poor ones. This also explains the reason for the location choices of large companies, which find social contexts less endowed with mobilisation capacity in the poorest areas (Oppio et al., 2015).

4.5 Conclusion

Both buildings energy insulation and photovoltaic show how the manner of building energy transition policies has reproduced existing social and territorial inequalities. They have probably been further aggravated by incentive policies, which have shifted wealth from the less affluent to the richer social classes. In the case of photovoltaics, many of the incentives for energy production even go beyond national borders in cases where foreign companies make speculative investments in Italy. The territorial dimension joins the social dimension. Both in the case of retrofits and in that of photovoltaics, there are territorial inequalities in the distribution of resources. In the case of retrofits, the inequality seems to be more due to the presence or absence in the territory of *organisational fields* capable of building economic chains around the renovation of homes and the availability of resources for initial investments. In the case of solar energy, however, the experiences of real energy colonisation are linked to the degree of marginalisation of some provinces. They represent places where global investors find it easier to penetrate with impactful projects on from a landscape and land use. According to several papers, these implementation dynamics of the energy transition weaken the possibility of proceeding faster. Even if there are no widespread collective actions opposing these policies, as is the case with the installation of wind turbines (see Chap. 2), they do not contribute to spreading the perception that the ecological transition brings distributed benefits to society and allows the living conditions of weaker social classes to be improved. In other words, they do not promote confidence in energy transition. If we look at the two cases addressed in this chapter, we are faced with two types of injustice: distributive and territorial. Distributive injustice concerns the distribution of the costs and benefits of the transition. In the case of the energy retrofit of homes, it is the wealthiest social classes that benefit from the incentive system. We can identify different types of benefits: firstly, even if indirectly, there is a redistribution of economic resources from the bottom up; secondly, housing disparities are growing as a result of the increase in the value of homes that have improved their energy class; thirdly, energy consumption, and therefore the costs of heating and electricity, are decreasing for social classes that do not have problems with energy poverty; finally, those who used tax deductions improved the healthiness of their homes with positive consequences on their quality of life. Spatial injustice, on the other hand, has to do with an unequal distribution of the benefits of the ecological transition in territorial terms. There are areas in which the transition becomes an element of democratisation of the energy system and areas, on the other hand, where the transition reproduces and accentuates a hierarchical and centralised model. In the first case, renewables become an instrument of social emancipation and citizen empowerment. In the second case, however, they favour new forms of dependence for the more fragile territories on the stronger ones. The costs of the transition are concentrated in some territories, while the benefits fall on much larger areas. Territorial injustice has been much debated in recent years and also provides an alternative reading key to the question of citizens' opposition even to renewable energy plants. Some authors underline that there is a link between the way to build incentive policies for the spread of renewables and the emergence of *negationist movements* on the climate and environmental issue. The so-called places left behind, where the energy transition is known in its form of energy colonisation, react by expressing a demand for social protection and a strong scepticism towards expert knowledge. The transition is not experienced as a tool of emancipation but as a device of the technocratic elites to concentrate power and wealth within a green rhetoric. The weakest areas, where this transition takes place, led by capital outside the territory, are the same areas where the wave of populism has spread in recent years, hostile in fact to the energy transition and to the recommendations of climatologists on the need to reduce climate-altering emissions. If it is true that political choices have led to an unequal and inclusive transition model with respect to social and territorial inequalities, it is possible to orient the energy transition in a new way. To do this, it is necessary that policies take on a *social and territorial posture*, working on a radical decentralisation of the energy system and using the energy transition to improve the quality of life of the weaker social classes. On these fronts, there are many micro-experiences that arise from civil society (Magnani & Osti, 2016), which place the dimension of the community at the centre of energy production and consumption. Starting from these experiences, which arise despite the obstacles of very rigid laws and regulatory frameworks, it is possible to outline new policies, which emerge from technological determinism and to recover the social dimensions of energy.

References

Baker, S., & Mehmood, A. (2015). Social innovation and the governance of sustainable places. *Local Environment, 20*, 321–334.

Barca, F., McCann, P., & Rodríguez-Pose, A. (2012). The case for regional development intervention: Place-based versus place-neutral approaches. *Journal of Regional Science, 52*, 134–152.

Blasutig G. (2017). Le funzioni intermediarie nella diffusione delle innovazioni. L'esperienza del fotovoltaico. *Rassegna Italiana Di Sociologia, 3*, 487–512.

Bouzarovski, S., & Simcock, N. (2017). Spatializing energy justice. *Energy Policy, 107*, 640–648.

Carrosio, G. (2015). Politiche e campi organizzativi della riqualificazione energetica degli edifici. *Sociologia Urbana e Rurale, 37*, 21–44.

Carrosio, G. (2020a). Resilience, cohesion policies and the socio-ecological crisis. In A. Balducci, D. Chiffi, & F. Curci (Eds.), *Risk and resilience* (Springer Briefs in Applied Sciences and Technology). Springer.

Carrosio, G. (2020b). *Povertà energetica: le politiche ambientali alla prova della giustizia sociale*, Working papers, Rivista online di Urban@it—2/2020

Carrosio, G., & Osti, G. (2011). *Un'analisi ecologica dell'economia civile del Nord Italia*. Aiccon Working Paper.

CRESME. (2017). Il recupero e la riqualificazione energetica del patrimonio edilizio: una stima dell'impatto delle misure di incentivazione. http://documenti.camera.it/leg17/dossier/pdf/Am0051d.pdf

Gottenhuber, S., & Mulholland, E. (2020). Implementing the sustainable development goals in times of rising right-wing populism in Europe. *Sustainability (Switzerland), 12*(20), 1–14.

Jenkins, K., McCauley, D., Heffron, R., Stephan, H., & Rehner, R. (2016). Energy justice: A conceptual review. *Energy Research and Social Science, 11*, 174–182.

Magnaghi, A. (2020). *Il principio territoriale*. Bollati Boringhieri.

Magnani, N., & Osti, G. (2016). Does civil society matter? Challenges and strategies of grassroots initiatives in Italy's energy transition. *Energy Research and Social Science, 13*, 148–157.

Matlock, A. S., & Lipsman, J. E. (2020). Mitigating environmental harm in urban planning: An ecological perspective. *Journal of Environmental Planning and Management, 63*(3), 568–584.

O'Sullivan, K., Golubchikov, O., & Mehmood, A. (2020). Uneven energy transitions: Understanding continued energy peripheralization in rural communities. *Energy Policy, 138*, 111–288.

Oppio, A., Corsi, S., Mattia, S., & Tosini, A. (2015). Exploring the relationship among local conflicts and territorial vulnerability: The case study of Lombardy region. *Land Use Policy, 43*, 239–247.

Sovacool, B. K., & Dworkin, M. H. (2015). Energy justice: Conceptual insights and practical applications. *Applied Energy, 142*, 435–444.

Spaargaren, G. (2000). Ecological modernization theory and domestic consumption. *Journal of Environmental Policy & Planning, 2*(4), 323–335.

Stavins, R. N. (2003). Experience with market-based environmental policy instruments. In *Handbook of environmental economics*. Elsevier Science.

5

Community Renewable Energies Between Social Enterprises, Social Movements and Hybrid Networks

Natalia Magnani

Abstract This chapter deals with the emergence from below of new forms of collective action that unite producers and consumers of renewable energies and intend to generate new exchange systems. These innovative forms of prosumption can be analysed through the construct of community renewable energy. This is a broad concept, comprising a variety of experiences concerning communities of interests or communities of place, which share the development of a project for the production of renewable energy and the economic and social benefits deriving from it. In the first part of the chapter, we will highlight the key theoretical perspectives that have been emerging in the literature around the concept of community energy. In the second part, we will focus on community energy in Italy.

Keywords Collective action • Community energy • Prosumers • Social enterprise • Social movement

---

N. Magnani (✉)
Department of Sociology and Social Research,
University of Trento, Trento, Italy
e-mail: natalia.magnani@unitn.it

N. Magnani, G. Carrosio, *Understanding the Energy Transition*,
https://doi.org/10.1007/978-3-030-83481-4_5

5.1 Community Renewable Energy: A Boundary Object

A growing body of literature (Wolsink, 2006; Jobert et al., 2007; Walker & Devine-Wright, 2008; Tarhan, 2015; Yildiz et al., 2015) points out that a greater direct involvement of society in the production and distribution of renewable energy sources (res) is a necessary condition if progress in the energy transition is to be made.

In this regard, Seyfang and Smith proposed the concept of *grassroots innovation* to describe "innovative networks of activists and organizations that develop bottom-up solutions for sustainable development; solutions that respond to the local situation and the interests and values of the communities involved" (2007, p. 585).

Grassroots innovations do not start from conventional market players and are not driven by purely economic reasons but rather by value options (Pellizzoni, 2018). Furthermore, contrary to ecological modernisation initiatives of traditional business, characterised by the prevalence of technological innovations, grassroots innovation projects tend to focus on the development of organisational innovations, the experimentation of new agreements and the creation of new social institutions (Doci et al., 2015).

There are many examples of grassroots innovation in the energy sector. Among them, a leading role both at the level of attention of academic research and of social diffusion is covered by collective projects for the management and production of res, which promote the idea of a public, active, technically competent and able to adopt an ethics of mutual help also on the energy issue.

The innovative nature of these experiences is related to the fact that they give rise to forms of management, production and distribution of renewable energy that diverge from traditional organisational models centred on the market or the state. There are several terms that are used in various countries: énergies partagées in France, bürgerenergie in Germany and community energy in the UK. These are all names used to indicate a new relationship between society and its energy system. Among these terms that of community energy or community renewable energy (CRE) is the most widespread in the international academic debate.

In an attempt to offer a definition of this diverse and dynamic domain, Walker and Devine-Wright (2008) stress the importance of jointly considering processes and results. They therefore propose to consider as CRE all those projects which, in varying degrees, are developed or managed through an open and participatory process and which produce local and collective benefits. Similar is the definition proposed by Bauwens et al. (2016), according to which CRE can be defined as all those initiatives activated by citizens, of a formal or informal type, which propose collaborative solutions on a local basis to facilitate the development of energy sustainability technologies.

As highlighted by Walker (2008), these projects can take diversified organisational and legal forms: from cooperatives owning energy infrastructures—the most widespread model—to non-profit entities that manage the plants on behalf of the local community, to co-ownership of green energy projects by communities, businesses and local governments.

CRE can function as energy producers, as suppliers or combine these areas with the promotion of energy saving projects. Moreover, as pointed out by Tarhan (2015) and Magnani and Osti (2016) they can be based on local communities or on communities of interests. The breadth of the definition of CRE, deriving from the complexity of the very concept of community, has been problematised by various authors (e.g. Hoffmann & High-Pippert, 2010; Hoffman et al., 2013; Seyfang et al., 2014). On the positive side, it was highlighted that the adoption by project proponents, and in part also by researchers, of a loose definition of CRE guarantees community actors the flexibility to develop initiatives that are adequate and sensitive to the needs and characteristics of local communities (Hicks & Ison, 2018; Parkhill et al., 2015).

On the negative side, it was underlined how a too broad definition of CRE risks losing sight of the principles of sustainable community development and energy democracy to which many projects aspire or even legitimising forms of "business as usual" types of development (Rogers et al., 2012).

To solve the inherent problematic nature of the concept of community, some authors (Bristow et al., 2012; Becker & Kunze, 2014; Kunze & Becker, 2015) propose replacing CRE with the analytical term collective and politically motivated energy projects (CPE). This term gives particular importance to the "political" element. Starting from a comparative analysis of some international case studies, Kunze and Becker (2015) believe that civil society initiatives on energy often have an agenda of aspirations that go beyond electricity generation and address a "politics of the possible", which seeks to use the limited room for manoeuvre to achieve a gradual change in the organisation of society. Key objectives that characterise these forms of collective action are a general reduction in energy consumption, the protection of biodiversity, more sustainable agriculture, greater social equity and the empowerment of disadvantaged groups.

If, in some cases, this dimension of transformation and political engagement can be identified, as we will see in the following paragraphs, however, it does not always occur. Therefore, it seems more useful and parsimonious to continue to use a more generic term such as CRE. In doing so, we agree with Pellizzoni's (2018) approach, according to which the term CRE can be considered a *boundary object*, that is a bridge concept, which, precisely because of its indefinite meaning, can facilitate contact and dialogue between subjects, scholars and diversified interests.

On the one hand, the sociological literature (e.g. Gubbins, 2007; Rogers et al., 2012; Seyfang et al., 2014; Tarhan, 2015; Yildiz et al., 2015) generally agrees in attributing economic, social and environmental advantages to community initiatives on renewable energy that distinguish them from governmental or company-led initiatives. On the other hand, problematising this kind of initiative on a more general level, Pellizzoni (2018) highlights that the growing popularity of the term "energy community", especially in the political discourse, points out the risk that these forms of participation of civil society, can be functional to the needs of capital rather than expressing instances of change, as they end up weakening political conflict.

5.2 Community Energy as Social Enterprise between Public and Mutual Benefits

If we analyse the literature on community energy in search of the dominant theoretical approaches, it is evident that we are faced with an emerging research field in which the theoretical references are still quite disparate and fragmentary. Even with these limitations, in the following we will highlight the emergence of theoretical perspectives that point out from time to time some characteristics and central dimensions of the phenomenon.

A first expanding line of research adopts a perspective that recognises community initiatives for the production and consumption of renewable energy—and especially cooperatives—as a form of social enterprise (Van der Horst, 2008; Huybrechts & Mertens, 2014; Kalkbrenner & Roosen, 2016; Bauwens & Defourny, 2017; Becker et al., 2017), a typical object of third sector studies. Indeed, as Gross and Mautz point out (2015, p. 100), most civil society initiatives on res are characterised by their hybrid structure. They combine the promotion of the common good with economic efficiency and with forms of mutual aid to their members. Their characteristic is precisely that of holding together apparently incongruent perspectives, thus managing to mobilise a specific combination of resources.

In particular, the literature that adopts this approach (Huybrechts & Mertens, 2014; Bauwens & Defourny, 2017) points out that one of the major contributions of social economy organisations in promoting sustainable transition is the creation and mobilisation of non-market resources like social capital. In the literature, social capital has been considered at the same time a necessary condition for the development of innovative organisational forms and a possible result of cooperative behaviour within organisations (Putnam, 1993).

The literature (e.g. Bauwens & Defourny, 2017) also highlights that the organisational structure adopted significantly influences the type and level of mobile social capital. In particular, significant differences can be identified between mutual benefit-oriented organisations, which basically aim to meet the needs of their members, and public benefit-oriented organisations, which aspire to enhance the well-being of a larger community or society as a whole.

Bauwens and Defourny (2017) interestingly try to understand how energy cooperatives position themselves with respect to this distinction and to what extent it is relevant to the type of social capital mobilised. By comparing two Belgian energy cooperatives, namely Ecopower (47,400 members) and BeauVent (2,400 members), they reach two kinds of conclusions. On the one hand, it is emphasised that often in the case of small energy cooperatives, which due to organisational and economic limitations can only deal with production and not with distribution, the traditional identification between members and users is not present. Mutuality will therefore also be weak. On the other hand, it is highlighted that while it is true that mutual benefit-oriented cooperatives are often characterised by a closed network consisting of strong ties, this relationship however is also mediated by two aspects: spatial factors and the structure of social interactions within the organisation itself.

With regard to the first factor, it is argued that proximity facilitates the activation of group social norms. As regards the second aspect, various studies have shown that social capital is negatively influenced by the growth of the structure of cooperative organisations. The greater organisational complexity implies that the number of members becomes larger and more diversified and individual trust and involvement decrease. This has clearly emerged in the case of Ecopower, which has undergone a rapid internal evolution in order to close the circle of production and distribution.

5.3 Community Energy as Strategic Action Field

Alongside the analysis of energy cooperatives as forms of social enterprise, a growing number of contributions analyse the dynamics underlying the emergence and affirmation of community energies by resorting to theoretical frameworks which can be traced back to the analysis of social movements.

This is in particular the case for the theory of strategic action fields. The variegated literature to which it refers (e.g. DiMaggio & Powell, 1983; Fligstein & McAdam, 2012) defines the field as a highly complex arena where actors who aim to preserve the existing balance (incumbent) and actors who want to modify it (challenger) confront each other. These two categories of actors possess opposite positions and disparate resources: the former is characterised by a disproportionate ability to influence the characteristics of the field, while the latter occupy niches in which an alternative vision is articulated. To produce change, challengers must be able to create common visions, generate shared identities and mobilise forms of collective action.

Mey and Diesendorf (2017) use this approach in order to analyse the emergence and evolution of wind farms in Denmark. They underline how in this country community energies can be considered a real field of strategic action with specific characteristics: first is the association with a specific technology for the production of renewable energy—wind power and second is the organisational structure of the property. In particular, it refers to the long Danish tradition of energy cooperatives, which originated in the cooperative movement in 1860. Finally, a third characteristic of community energies as a strategic field of action is the regulatory dimension.

The strategic action fields approach identifies the process of establishing a field with the commitment to the institutionalisation of its principles. In Denmark, the field of community energies was solidified when groups of actors with specific social skills coordinated for the implementation of alternative energy scenarios through the creation of associative structures, which were developed also thanks to a favourable regulatory framework.

Bomberg and McEwen (2012) also analyse community energies as a social movement. In their work, they refer to a modified resource mobilisation framework, which focuses on the normative and more subjective dimension of mobilisation. In particular, the centrality of two types of strategic symbolic resources for the development of community energies is emphasised: collective identity and the search for autonomy. The empirical research conducted by the authors highlights that often community energy projects concerning res emerge not so much on the basis of environmental concerns but as a means of overcoming dependence on the outside, to regain greater control over costs and energy supply, and also more generally over local development trajectories. This is true both for remote rural communities already accustomed to a certain degree of autonomy and for deprived urban communities aiming to generate clean energy and income independently of and in opposition to government.

5.4 Community Energy and Socio-ecological Systems

The studies examined above underline how the use of symbolic resources is central to the effectiveness of forms of collective action and mobilisation regarding energy. However, what seems to be lost sight of in these studies is the context in relation to which these symbolic resources develop.

An integration in this direction is instead offered by another approach, which analyses energy communities, in particular cooperatives, as a result of the interaction that takes place within complex socio-ecological systems (see Bauwens et al., 2016).

The focus of this perspective is on the different contextual variables influencing the situation in which a multiplicity of individual or collective actors interacts with each other. In particular, reference is made to the resource system and the governance system. The former identifies the biophysical and technical systems from which resources are extracted. The latter concerns the set of processes, institutions and rules that shape the behaviour of the actors.

Bauwens et al. (2016) use this framework to compare the context for the development of cooperative initiatives on renewable energies in Germany and the UK. Germany, like other Northern European countries, is characterised by a strong tradition of energy activism linked to the anti-nuclear movement. However, Bauwens et al. (2016) underline that the recent re-emergence of the cooperative model is to be interpreted as a consequence of the economic crisis and the search for new organisational models. In this framework, expectations related to the profitability of investments have played an increasing role. As a result, the social base of the pro-renewable coalition has expanded far beyond the confines of the anti-nuclear movement to include professionals, farmers and small businesses. This has also been favoured by a legal framework that not only supports decentralised and small-scale forms of renewable electricity generation but also supports third sector organisations in creating a solid financial base (e.g. through incentive tariffs, tax exemptions, loans).

As for the UK, although it can be considered one of the best sites for wind energy, the cooperative sector is underdeveloped when compared with Germany. In the UK, the initial development of res was marked by unfavourable conditions for civil society initiatives, especially concerning the governance system. In particular, three negative factors are underlined: the lack of a tradition of cooperative organisations, the absence of a strong energy movement and an energy policy in favour of large-scale plants owned by large private companies. Furthermore, the UK does not have a specific law on cooperatives, to the point that any form of business can be structured around cooperative principles.

5.5 Community Energy as a Localised Socio-technical Translation Process

In addition to the theoretical approaches analysed above, some recent studies developed a perspective based on the actor-network theory (ANT) and the sociology of translation. Several works have underlined the relevance of these theories in analysing issues of environmental sociology, in particular the diffusion of innovations and how socio-technical dynamics work.

With regard to community energies, the work by Scotti and Minervini (2017) is particularly interesting because it draws on the sociology of translation to analyse the way in which national policies for sustainable energy are translated at the local level. It is emphasised that the energy transition can be conceptualised as a multiple and heterogeneous action programme, enacted by a plurality of entities—human and non-human.

The purpose of the analysis is to trace the translation process that connects the national and supranational regulatory framework with the agency of objects, actors, institutions and discourses implemented at the local level. This process is considered performative in the sense that the multilevel network, which is the object of sociological analysis, is built through the mutual influence and interaction between multiple elements.

This framework is applied by Scotti and Minervini (2017) to a small case of a rural town in Southern Italy. Faced with the challenges of depopulation and unemployment, initiatives regarding sustainability have emerged, focused on the construction of two municipal photovoltaic plants and a wind turbine, through the articulation of a network between the municipality, regional offices and environmental associations. In assembling this network, there emerges the importance of "focal actors" (Callon, 1986), which are not essential in themselves but become crucial in a specific space and time when their interests and actions are able to produce a synergy that binds other entities together through the construction of an obligatory passage point.

In the case study analysed by the authors, the focal actor, who activated the network, is identified with an engineer who worked in the mobility sector and who was a member of the municipal council. Thanks to his "bridge" position, he was able to introduce the issues of energy sustainability to the political agenda. His technical expertise was extremely relevant in prefiguring energy initiatives in terms of the action programme. The problematisation articulated by this bridging actor was also effective in connecting the local project to supralocal resources and initiatives, such as national incentives and funding from the European regional development fund.

In this perspective, renewable energy innovation, rather than a simple result of charismatic personalities, is to be considered as a process through which actors negotiate identity, rationality and relationships with other human and non-human actors, which allow the emergence of a new socio-technical assemblage.

5.6 Forms of Collective Organisation of Renewable Energy in Italy

Most of the research on community energies so far has concerned Northern European countries. Many authors point out that studies of community initiatives concerning the countries of Southern Europe are lacking. Indeed, the literature explicitly points out the absence of such forms of democratic participation in the production-consumption of energy and attributes it mainly to three factors: a lower disposable income per capita, a lower diffusion of renewable energies and a lower diffusion of environmental values (Schreuer & Weismeier-Sammer, 2010; Huybrechts & Mertens, 2014; Romero-Rubio & Díaz, 2015; Pellicer-Sifres et al., 2018).

Recently some studies have, however, begun to map this still fragile and fragmented field, even in some countries of Southern Europe, with the aim of better understanding its differences and weaknesses. As in the previous chapters, in this part we focus on the emerging empirical research concerning community energies specifically in the Italian case (Magnani & Osti, 2016; Candelise & Ruggeri, 2020).

These studies highlight that it is only in recent years—starting from the second half of the 2000s—that innovative experiences of "energy civism" have emerged in Italy regarding the consumption, production and, to a limited extent, the distribution of renewable energy. These new forms of participation and organisation have largely involved solar photovoltaics and to a lesser extent wind power. This is mainly attributable to the generous incentives enjoyed by photovoltaics from 2005 to 2013 (Antonelli & Desideri, 2014), to the decreasing cost of this technology in relation to the possible gains as well as the availability of plant engineering know-how in the area (Candelise et al., 2013).

The participation of citizens in the energy market in Italy essentially gives rise to three new forms of social relations: the creation of purchasing groups for photovoltaic panels and green energy, the development of forms of collective management of production plants through the creation of cooperatives of prosumers and new forms of re-intermediation through the creation of secondary cooperatives for the management of the distribution and sale of electricity produced by collective plants. In the following we analyse the characteristics of each of them and in the Conclusion, we will highlight the relevance of some of the theoretical approaches considered above to understand the limits of community energy development in Italy.

5.6.1 Solidarity Purchasing Groups of Photovoltaics

The first form of citizen involvement in energy democracy follows the forms of mobilisation and self-organisation that have developed in the *food supply chain* through solidarity purchasing groups (Fonte, 2013). As in the food sector, green energy consumer associations have also emerged in the energy sector since about 2000. The most important of these, Energoclub, has taken the form of a large purchasing group aimed at supporting citizens in the choice and installation of domestic photovoltaic solar systems or condominium systems. Five other smaller associations operate in Northern and Central Italy. The members of these associations are single families or more often small purchasing groups of photovoltaic systems (Gruppo di acquisto solidale fotovoltaico, Gasfv). The latter are made up of consumer groups, sometimes created ad hoc, sometimes already operating in the food chain, whose goal is to generate a different economic model, based on a direct relationship between consumers and producers.

The motivation for joining a Gasfv is mainly to purchase a product—in this case a photovoltaic system—at a lower cost thanks to the wholesale purchase. However, the advantages also concern other aspects. First of all, the construction of a photovoltaic system requires a certain level of technical competence in the choice of technology and suppliers. The role of a Gasfv therefore becomes that of setting a price that is adequate to the quality standard required by the members of the purchasing group. The main features required concern, for example, the origin of the modules, the transparency of the supplier with regard to taxation and work, the technical guarantee of the systems, the maintenance offer and safety at work during installation. The experience gained by a Gasfv allows the most efficient and safe products to be chosen. Moreover, European or national products are often preferred in order to promote local development and reduce the international movement of goods.

The Gasfv also offers a support network for solving various problems related to the installation, from technical problems to administrative procedures linked to incentives, to relations with the manager of the energy system rather than with the manager of the local network.

Joining a Gasfv also offers the possibility to compare the production data of one's own plant with those of the other members; thus, it can result in greater awareness regarding energy behaviour and energy saving.

Finally, the associative advantages related to the price and quality of goods concern not only the technology but, albeit to a lesser extent, also the flow of energy. In fact, some of these groups of consumers also promote agreements for the collective purchase of electricity from utilities that certify the green quality of their sources.

In general, Gasfvs are a more widespread form of citizen participation than cooperatives. They are based on large networks, focused on building a community of interests rather than a community of place. Despite the advantages we have seen at an individual or small group level, however, the possibility of having an impact on the energy transition for this type of collective action seems to be limited to the possible indirect effects produced on the supply of photovoltaic technology. This is proved by the fact that once the photovoltaic incentive was over, these groups almost completely disappeared, without offering themselves for other products or services of the energy transition.

5.6.2 Cooperatives for the Production and Consumption of Renewable Energy

A greater degree of participation in energy democracy is evident in the creation of cooperatives that unite in a single entity production and consumption of renewable energy. The involvement of local communities in the production and management of renewable energy through cooperative forms is not new in Italy. However, until recent years, it has been largely limited to a single geographical area, identified with the Alpine area. The so-called historical cooperatives are hydroelectric cooperatives that arose in connection with the construction of medium-large dams in the first half of the twentieth century in the Alps (Mori, 2014). The goal of these projects was to generate development and well-being for the people of the mountain areas through the production and distribution of electricity. Most of the historic cooperatives still exist and are located in Trentino-Alto Adige. By virtue of their social function, even after the nationalisation of the electricity grid (1962), they were granted a special regime in derogation from the general rules of the electricity State monopoly. This allowed them to retain ownership of the local electricity grid and to distribute the electricity they produced.

Alongside the example of historic hydroelectric cooperatives, it should also be noted that, as already mentioned in Chap. 2, in South Tyrol there is a strong tradition of community projects (cooperatives) for the production of energy from biomass (wood), which in many cases has resulted in the creation of district heating systems that guarantee the self-sufficiency of the local community from the point of view of thermal energy (Wirth, 2014). However, these initiatives remain exceptional in the Italian framework since they are largely explained by the presence of historically rooted institutionalised networks based on a strong territorial and ethnic identification.

In the last ten years, innovative community initiatives for the shared installation of res plants, in particular of "collective photovoltaics", have emerged. In most cases, these experiences have taken the form of cooperatives with prevalent mutuality (art. 2512 of the Italian Civil Code). The primary purpose of the cooperative is the supply to its members of goods and services at prices defined by the members themselves. Due to the higher ultimate purpose of mutuality, these cooperatives enjoy an advantageous tax regime.

As highlighted by the empirical research by Magnani and Osti (2016), this abstract model then materialised in diversified forms of organisation, resulting from the interaction between geographical characteristics, local institutional system—formal and informal—and motivational actions of the leaders who initiated the project.

In particular, we can identify three types of cooperative organisations on renewable energy. First of all, we find those cooperatives that have emerged from the galaxy of solidarity purchasing groups and solidarity economy districts, thus within a subculture strongly oriented to the themes of participation, self-management, solidarity and environmental sustainability. For example, this is the case of the coop. Retenergie, which arose from the desire, on the part of a solidarity purchasing group of Piedmont, to share the energy produced from renewable sources. In 2018, in the attempt to close the energy production-consumption cycle it has merged with consumer cooperative ènostra (see Sect. 5.6.3).

The cooperative gathered about 1100 members with a presence in 15 Italian regions. It built nine solar plants for a total power of approximately 630 kilowatt-peak (kWp). The members contributed with variable amounts to the cooperative's share capital. An annual average remuneration was recognised on the social loan, which was around 1.5%–3% of the paid-up capital. Most of the plants were built on public buildings. The agreement with the public administration provided that the latter will be offered the possibility of receiving a surface that is right for the use of the roof, a small-scale plant as compensation or the elimination of asbestos from the roof cover.

This type of cooperative was characterised by a very complex participatory governance system. The cooperation developed through the opening of new territorial nodes. Once a new opportunity or the will of a group of people had been identified, a so-called local node was set up, usually made up of about thirty members. The territorial node, to which the construction of a specific plant can be associated, organised its own structure and activity but always acted within the cooperative. The Retenergie cooperative aimed to maximise the environmental and social benefits of renewable energies through the creation of small-scale collective plants, without any economic risk on the part of the citizens.

Other solar cooperatives that have recently emerged are characterised by a more entrepreneurial vocation. They have developed on the impulse of specific focal actors, that is new ecopreneurs who try to combine environmental awareness with their commercial activities in an attempt to promote a transformation of the economy in a more environmentally friendly direction (Gibbs, 2009, p. 65).

Taking up a definition previously analysed, while Retenergie is characterised by an attempt to turn to public benefit, these cooperatives are more focused on a mutual benefit approach.

This is the case, for example, of the Energyland project, born in the province of Verona on the initiative of a group of entrepreneurs with work experience as a manager of a local energy utility. The Energyland photovoltaic cooperative is the last stage of a process in which the initiators have mobilised a vast network of economic and financial actors (banks, insurance companies, companies) for the creation of a collective solar plant of just under 1 MW. The plant was built on three hectares of land, previously used for sheep grazing.

In this case, the members of the cooperative were largely recruited once the plant was completed through a vast advertising campaign conducted in the local press and on the web. To become a member of the cooperative, the purchase of one or more shares of the plant, equivalent to 1 kW (up to a maximum of 6 kW), was required. The cost was set at 3600 euros per kW.

Against the considerable initial investment required to join the collective solar project, a member of the Energyland cooperative was able to obtain a return on invested capital ranging from 6.5% to 8.8% (Candelise & Ruggieri, 2020). The Energyland cooperative then acted as an intermediary for the creation of two other production cooperatives. In 2011, thanks to the surplus obtained from the sale of energy to the grid, the initial group of entrepreneurs purchased an abandoned area in the province of Lecce, on which they built a new photovoltaic plant (9 MW). The WeForGreen cooperative was established, aggregating around 190 members from 17 Italian regions. Finally, in 2016, in the same province, on the basis of the same financial network, and with the same development strategy, a new plot of land was acquired, and the WeForGreen Sharing cooperative was born, aggregating 104 families.

All three cooperatives are in fact coordinated by the For-Green company, responsible for the development of photovoltaic systems and energy services and directed by the founding ecopreneurs. The basic idea of this initiative is to use the cooperative form to share environmental services beyond mere production. To do this, cooperatives aim to integrate corporate participation with consultancy in the field of energy saving and with the sale of ethically characterised services (concerning, e.g. the real-time monitoring of energy consumption).

As pointed out by various authors (Magnani et al., 2018; Candelise & Ruggieri, 2020), although the cases considered so far fall within the definition of "energy cooperatives", it is clear how the different organisational and relational cultures from which they emerge, as well as the views of their initiators, imply significant differences in their financial and governance structure. In particular, it should be emphasised that on the one hand, entrepreneurial initiatives are characterised by greater certainty and relevance (also quantitative) in relation to the economic profit deriving for each shareholder from the production of renewable energy. However, this is accompanied by limited citizen participation in the ordinary administration of each plant. On the other hand, a project like Retenergie, in the face of extremely limited or non-existent economic returns for members, was characterised by greater attention given to participatory governance through the organisation of a nodular structure that aims to promote direct access to citizens also in relation to strategic decisions regarding the cooperative.

Despite these differences, the cooperative forms analysed above tend to be structured mainly as a "community of interests", aimed at attracting a broad public of citizens united by sharing an environmental interest or rather an economic one that is linked to the reduction of costs. The spatial contiguity between production and consumption is therefore neither a prerequisite nor a prevalent feature.

The research, however, reveals another type of energy cooperative, characterised to a greater extent in a local sense and which takes place within small rural municipalities (see Magnani et al., 2017). These are innovations by territorial ecopreneurs who establish themselves within a "local community" and who leverage relational and trust resources linked to belonging to the same territory. This type of renewable energy cooperative is conceived as a tool aimed to recover economic resources to stem processes of marginalisation and possibly trigger virtuous local development trends. For example, this type includes the So.Le cooperative developed in Ledro, a mountain town of about 5000 inhabitants in an Alpine valley, from a meeting between an entrepreneur who owned a cooperative company specialising in the installation of photovoltaics and some local actors involved with public institutions and sensitive to environmental issues. The goal was to make photovoltaic technology the basis of new mutual relations and a tool to generate development opportunities for a fragile area subject to depopulation processes. Central to the project was the building of two collective solar systems: one on a public structure, the other on a private sawmill, for a total of 98 kW installed. Approximately one hundred shareholders currently participate in the ownership and management of the plants without any financial commitment, except for the signing of a pro quota surety to guarantee the investment in photovoltaic panels to be made at the local bank. The cooperative was thus able to ensure its members a saving of approximately 25% on electricity bills. Furthermore, as a secondary effect, new jobs have been created locally in the sector of photovoltaic installations, and a series of initiatives for the protection and enhancement of the territory for tourism purposes, such as the recovery and maintenance of old trails, were implemented.

5.6.3 Cooperatives for the Sale and Distribution of Renewable Energy

The experiences of energy cooperatives just presented, although diversified from each other, try to promote an alternative model to the conventional system of production and use of energy; however, in doing so, they clash first of all with obstacles linked to the unbundling processes.

The liberalisation of the energy sector has meant the opening of some (limited) opportunities for new players in energy production, but it has also introduced new rules at the European level, which, in order to regulate competition, require the separation between production, marketing and supply of electricity. In fact, the existing legislation provides that the electricity produced by renewable plants, also depending on the type of plant and the type of incentive they receive, must be sold only to wholesalers, other energy sales companies or must be withdrawn by the energy services operator (Candelise & Ruggieri, 2020). As in other European cases, so too in Italy this has created great difficulties for non-historical energy cooperatives. In particular, it has made it difficult, if not impossible, to implement the mutual purpose.

Cooperatives, unable to supply electricity directly to their members, had to invent new organisational forms to mend the prevailing mutuality criterion at the basis of their projects. The simplest way for most of the cooperatives was initially to identify a medium-sized utility interested in financing the social economy and make it a member of the cooperative itself. In most cases, the choice fell on Trenta S.p.A., a company operating in the province of Trento and able to offer 100% green energy produced by hydroelectric plants present in the Alpine areas of Northern Italy. This form of traditional commercial intermediation has been considered by most of the energy cooperatives as a minor and transient evil, necessary for a first phase of the economic stabilisation of the sector.

Innovative attempts by the third sector have recently been implemented to reconstruct the original mutuality through the creation of new cooperatives to support the former. The most promising attempt seems to be the establishment of the cooperative ènostra. In 2014, Retenergie, EnergoClub and some experts from a sustainable business incubator joined together to create this new cooperative aimed at the sale of renewable energy. Based on similar experiences in other European countries, such as that of the Belgian cooperative Ecopower, the idea was that the ènostra cooperative could buy electricity directly from the plants owned by the members who helped to create it.

The attempt was to recreate, within the traditional electricity market, a cooperative energy network that brings together users and producers, excluding the intermediation of medium and large traditional utilities. The final goal promoted was to disintermediate at least one niche in the energy market. In so doing the ènostra initiative proved successful. Indeed, when ènostra started operating and selling energy, in 2016, Retenergie transferred the energy of its plants that did not have a tariff to the cooperative comprehensive incentive, ending the agreement with the utility Trenta spa. In this way ènostra could sell as per its mission cooperative energy to its own members and to those of Retenergie.

The intention was to close the circle between production and consumption as is already the case for many energy cooperatives in Europe (e.g. the Spanish Somenergia or the French Enercoop), which were born from social movements similar to that of Retenergie.

However, the attempt to close the production-consumption circle in a regulatory context, which was not favourable to energy cooperation and development of short production and consumption chains, proved also difficult from the point of view of communication with consumers. For the members of Retenergie the fact that in order to purchase energy from their cooperative it was necessary to become members of another cooperative was not immediately understandable. Thus, at the beginning of 2018 as a result of a joint general assembly the merging two cooperatives was decided. E'nostra is currently the first and only initiative in Italy that includes both the production and sale of renewable energy in the same cooperative organisation.

The cooperative model of ènostra is based on a widespread shareholding structure with a low entry threshold (a basic investment of 50 euros is required to become a member). The challenge of the project is to expand the number of members as much as possible. The difficulties encountered are due to the problems of scaling the action aimed at exerting pressure on the transformation of the dominant system. As we have also seen for other European cases, this represents a crucial challenge for energy cooperatives because it requires a radical transformation of their relational structure. It also has consequences at the level of collective identity and values. In particular, organisational growth requires reshaping the relationship with the territory and local communities and at the same time being able to keep together different forms of consumerism.

5.7 Conclusion

As this chapter has shown, in Italy, a number of emerging community energy initiatives can be detected. From an analytical point of view, they can be placed at different points on the representative axis illustrating the continuum between *communities of interest* vs *communities of place* on one side, and on the axis of mutual benefits vs public benefits on the other. In fact, in all cases of cooperation on energy, there is an ambivalence constant between mutual benefit objectives vs public benefit on the one hand and a concomitant tension between relational/spatial forms more focused on the local community vs the community of interests on the other. The first aspect concerns the tension between an organisational structure that substantially aims to respond to the needs of its members and a structure oriented instead to enhance the well-being of a larger community or society as a whole.

The second aspect concerns the spatial relational link and in particular the proximity, more or less close between the territory, whose resources are used for energy production, and the community of energy prosumers. The two extremes in this case consist of the community of place, in which the owners of the plants coincide with the citizens who live in the territory in which they are localised, and of the community of interests, aimed instead at attracting an audience of a large number of citizens without strict ties of local belonging, based on sharing an environmental or civic interest rather than an economic one.

If we try to apply the theoretical frameworks analysed in the first part of the chapter, we can stress that community energy in Italy struggles to be read according to the interpretative scheme of the strategic action field. The innovative experiences, albeit present, compared to other European countries remain limited in number and in the capacity to take hold of society. This is evident from the small number of members that make up the cooperatives analysed above and from the difficulty they have in expanding from an organisational point of view, especially in the distribution and sales segment. Furthermore, the initiatives examined above do not constitute a coherent movement, but the heterogeneity and fragmentation of the aims and organisational forms prevail. This happens despite the fact that, as in countries of Northern Europe, Italy has also experienced very strong growth in the production of electricity from res and has adopted a very generous incentive policy.

The experiences of existing community energies in Italy and their weaknesses can be read as the result of the interaction between a multiplicity of factors that emerged in the theoretical approaches explored above: regulation and institutional set-up, ecological and social characteristics of the context and focal actors.

First of all, there are constraints related to the organisation and regulation of the national energy system. As regards the organisation of the electricity market, the reference to municipal socialism is particularly relevant (Dogliani, 2002). This is a movement that developed in Italy in the last century, aimed at guaranteeing access to local public services for the entire population. It led to the nationalisation of the electricity grid in

1962. In 1999, the liberalisation of production and distribution took place. However, the grids remained a natural monopoly, and in the other segments (production and distribution) the Italian electricity market was also characterised by an oligopoly of a few operators (Enel, Eni, Edison, Sorgenia and some municipal companies) with strong participation of public bodies.

This intricate and nebulous organisational interweaving is combined with rigid institutional anti-competition rules that impose a separation between production, marketing and the supply of electricity, which strongly affects the possibilities of entry for new players in the energy system.

To these factors, one must add the importance of the prevailing forms of capitalism. In Italy, accumulation in the energy system is marked by a sort of duopoly between the state and large companies. The old centralised and integrated energy system, based on large plants and extensive networks, required huge investments, which only actors with a good storage capacity were able to make. The state and large companies were therefore the only actors able to intervene in the energy sector with powerful technical and financial means.

Furthermore, the Italian economic system is characterised by a strong north-south territorial polarisation between natural assets and capital. While res are mainly concentrated in the centre-south of the country, widespread capital, large industry, technology and investment capabilities are concentrated in the North. The result of this dualism and the consequences of the power structures here highlighted are similar for photovoltaic and wind power. As regards the first, an individualistic pattern with more than 500,000 citizens with an individual/family system, mainly concentrated in the Northern regions, was accompanied by the installation in the South of the country of large ground systems by few large private companies. As regards wind power, in the face of a very limited development of mini-plants, we are witnessing the prevalence of a form of *colonisation* of the South by large plants owned by private corporates, without any participation of the public.

As pointed out by Magnani and Osti (2016), in this situation, small cooperative enterprises and organised prosumers do not find it easy to operate. This is also linked to the fact that while in the countries of

Northern Europe, the political will to support the cooperative model has produced significant reforms and simplifications of the regulatory framework, this has not happened in Italy until now. It remains to be seen if the implementation of the directive 2001/2018/UE on community energies and collective forms of self-consumption will open some new opportunities for community energy also in Italy.

References

Antonelli, M., & Desideri, U. (2014). The doping effect of Italian feed-in tariffs on the PV market. *Energy Policy, 67*, 583–594.

Bauwens, T., & Defourny, J. (2017). Social capital and mutual versus public benefit: The case of renewable energy cooperatives. *Annals of Public and Cooperative Economics, 88*(2), 203–232.

Bauwens, T., Gotchev, B., & Holstenkamp, L. (2016). What drives the development of community energy in Europe? The case of wind power cooperatives. *Energy Research & Social Science, 13*, 136–147.

Becker, S., & Kunze, C. (2014), Transcending community energy: Collective and politically motivated projects in renewable energy (CPE) across Europe. *People, Place & Policy Online, 8*(3), 180–191.

Becker, S., Kunze, C., & Vancea, M. (2017). Community energy and social entrepreneurship: Addressing purpose, organisation and embeddedness of renewable energy projects. *Journal of Cleaner Production, 147*, 25–36.

Bomberg, E., & McEwen, N. (2012). Mobilizing community energy. *Energy Policy, 51*, 435–444.

Bristow, G., Cowell, R., & Munday, M. (2012). Windfalls for whom? The evolving notion of 'community' in community benefit provisions from wind farms. *Geoforum, 43*(6), 1108–1120.

Callon, M. (1986). Some elements of a sociology of translation: Domestication of the scallops and the fishermen of St. Brieuc Bay. In J. Law (a cura di), *Power, action and belief: A new sociology of knowledge?* (pp. 196–233) Routledge Kegan Paul.

Candelise, C., & Ruggieri, G. (2020). Status and evolution of the community energy sector in Italy. *Energies, 13*(8), 1888.

Candelise, C., Winskel, M., & Gross, R. (2013). The dynamics of solar PV costs and prices as a challenge for technology forecasting. *Renewable and Sustainable Energy Reviews, 26*, 96–107.

DiMaggio, P. J., & Powell, W. W. (1983). The iron cage revisited: Institutional isomorphism and collective rationality in organizational fields. *American Sociological Review, 48*(2), 147–160.

Doci, G., Vasileiadou, E., & Petersen, A. C. (2015). Exploring the transition potential of renewable energy communities. *Futures, 66*, 85–95.

Dogliani, P. (2002). European municipalism in the first half of the twentieth century: The socialist network. *Contemporary European History, 11*, 573–596.

Fligstein, N., & McAdam, D. (2012). *A theory of fields*. Oxford University Press.

Fonte, M. (2013). Food consumption as social practice: Solidarity purchasing groups in Rome, Italy. *Journal of Rural Studies, 32*, 230–239.

Gibbs, D. (2009). Sustainability entrepreneurs, ecopreneurs and the development of a sustainable economy. *Greener Management International, 55*, 63–78.

Gross, M., & Mautz, R. (2015). *Renewable energies*. Routledge.

Gubbins, N. (2007). Community energy in practice. *Local Economy, 22*(1), 80–84.

Hicks, J., & Ison, N. (2018). An exploration of the boundaries of 'community' in community renewable energy projects: Navigating between motivations and context. *Energy Policy, 113*, 523–534.

Hoffman, S. M., Fudge, S., Pawlisch, L., High-Pippert, A., Peters, M., & Haskard, J. (2013). Public values and community energy: Lessons from the US and UK. *Sustainability, 5*(4), 1747–1763.

Hoffmann, S. M., & High-Pippert, A. (2010). From private lives to collective action: Recruitment and participation incentives for a community energy program. *Energy Policy, 38*, 7567–7574.

Huybrechts, B., & Mertens, S. (2014). The relevance of the cooperative model in the field of renewable energy. *Annals of Public and Cooperative Economics, 85*(2), 193–212.

Jobert, A., Laborgne, P., & Mimler, S. (2007). Local acceptance of wind energy: Factors of success identified in French and German case studies. *Energy Policy, 35*(5), 2751–2760.

Kalkbrenner, B. J., & Roosen, J. (2016). Citizens' willingness to participate in local renewable energy projects: The role of community and trust in Germany. *Energy Research & Social Science, 13*, 60–70.

Kunze, C., & Becker, S. (2015). Collective ownership in renewable energy and opportunities for sustainable edgrowth. *Sustainability Science, 10*(3), 425–437.

Magnani, N., Maretti, M., Salvatore, R., & Scotti, I. (2017). Ecopreneurs, rural development and alternative socio-technical arrangements for community renewable energy. *Journal of Rural Studies, 52*, 33–41.

Magnani, N., Minervini, D., & Scotti, I. (2018). Understanding energy commons. Polycentricity, translation and intermediation. *Rassegna Italiana di Sociologia, 59*(2), 343–370.

Magnani, N., & Osti, G. (2016). Does civil society matter? Challenges and strategies of grassroots initiatives in Italy's energy transition. *Energy Research & Social Science, 13*, 148–157.

Mey, F., & Diesendorf, M. (2017). Who owns an energy transition? Strategic action fields and community wind energy in Denmark. *Energy Research & Social Science, 35*, 108–117.

Mori, P. A. (2014). Community and cooperation: The evolution of cooperatives towards new models of citizens' democratic participation in public services provision. *Annals of Public and Cooperative Economics, 85*(3), 327–352.

Parkhill, K. A., Shirani, F., Butler, C., Henwood, K. L., Groves, C., & Pidgeon, N. F. (2015). 'We are a community [but] that takes a certain amount of energy': Exploring shared visions, social action, and resilience in place-based community-led energy initiatives. *Environmental Science & Policy, 53*, 60–69.

Pellicer-Sifres, V., Belda-Miquel, S., Cuesta-Fernandez, I., & Boni, A. (2018). Learning, transformative action, and grassroots innovation: Insights from the Spanish energy cooperative Som Energia. *Energy Research & Social Science, 42*, 100–111.

Pellizzoni, L. (2018). Energia di comunità. Una ricognizione critica della letteratura. In G. Osti, L. Pellizzoni (a cura di), *Energia e innovazione tra flussi globali e circuiti locali*. Edizioni Università di Trieste.

Putnam, R. D. (1993). *Making democracy work: Civic traditions in modern Italy*. Princeton University Press.

Rogers, J. C., Simmons, E. A., Convery, I., & Weatherall, A. (2012). Social impacts of community renewable energy projects: Findings from a woodfuel case study. *Energy Policy, 42*, 239–247.

Romero-Rubio, C., & de Andrés Díaz, J. R. (2015). Sustainable energy communities: A study contrasting Spain and Germany. *Energy Policy, 85*, 397–409.

Schreuer, A., & Weismeier-Sammer, D. (2010). *Energy cooperatives and local ownership in the field of renewable energy technologies: A literature review*. Vienna University of Economics and Business.

Scotti, I., & Minervini, D. (2017). Performative connections: Translating sustainable energy transition by local communities. *Innovation: The European Journal of Social Science Research, 30*(3), 350–364.

Seyfang, G., Hielscher, S., Hargreaves, T., Martiskainen, M., & Smith, A. (2014). A grassroots sustainable energy niche? Reflections on community energy in the UK. *Environmental Innovation and Societal Transitions, 13*, 21–44.

Seyfang, G., & Smith, A. (2007). Grassroots innovations for sustainable development: Towards a new research and policy agenda. *Environmental Politics, 16*(4), 584–603.

Tarhan, M. (2015). Renewable energy cooperatives: A review of demonstrated impacts and limitations. *Journal of Entrepreneurial and Organizational Diversity, 4*(1), 104–120.

Van der Horst, D. (2008). Social enterprise and renewable energy: Emerging initiatives and communities of practice. *Social Enterprise Journal, 4*(3), 171–185.

Walker, G. (2008). What are the barriers and incentives for community-owned means of energy production and use? *Energy Policy, 36*(12), 4401–4405.

Walker, G., & Devine-Wright, P. (2008). Community renewable energy: What should it mean? *Energy Policy, 36*(2), 497–500.

Wirth, S. (2014). Communities matter: Institutional preconditions for community renewable energy. *Energy Policy, 70*, 236–246.

Wolsink, M. (2006). Invalid theory impedes our understanding: A critique on the persistence of the language of NIMBY. *Transactions of the Institute of British Geographers, 31*(1), 85–91.

Yildiz, Ö., Rommel, J., Debor, S., Holstenkamp, L., Mey, F., Müller, J. R., & Rognli, J. (2015). Renewable energy cooperatives as gatekeepers or facilitators? Recent developments in Germany and a multidisciplinary research agenda. *Energy Research & Social Science, 6*, 59–73.

6

Some Lessons on Energy Transition Learned from the Italian Case

Giovanni Carrosio

Abstract We are in a transition phase from an energy system based on fossil resources and centralised production plants to an energy system based on renewable sources and increasingly decentralised production plants. If for many years the sociological dimensions relating to the production and consumption of energy have been in the background, contemporary changes put sociological knowledge at the centre. This chapter highlights the new challenges of the social sciences in the energy transition.

Keywords Energy transition • Social sciences • Reticularisation • Decentralisation • Community

This book has highlighted some relevant aspects of the energy question as it presents itself in contemporary society. We are in fact in a transition phase from an energy system based on fossil resources and centralised production plants to an energy system based on renewable sources and increasingly

---

G. Carrosio (✉)
Department of Social and Political Sciences, University of Trieste, Trieste, Italy
e-mail: gcarrosio@units.it

N. Magnani, G. Carrosio, *Understanding the Energy Transition*,
https://doi.org/10.1007/978-3-030-83481-4_6

decentralised production plants. In this epochal passage, there are elements of continuity and discontinuity. Energy decentralisation meets other forms of reticularisation of the economy and society and can become a factor for a further push towards a new model. If for many years the sociological dimensions relating to the production and consumption of energy have been in the background, contemporary changes put sociological knowledge at the centre. The transition, in fact, lends itself to various social analyses, in particular on account of the fact that it represents a dynamic of social change, where technical innovations and societal change feed each other (Carrosio & Scotti, 2019). Within this change, many balances are modified, which in the past had stabilised. For this reason, sociology also reappears as a privileged discipline in investigating social change. The position that our discipline assumes today, however, is different than in the past. It deals more with energy issues through the theoretical contribution of the different schools of thought that distinguish the discipline rather than working on energy with the aim of re-establishing its theoretical assumptions. As we stated in Chap. 2, there is no new socio-energetic, but there are many contributions that investigate the social aspects which intervene in the conflictual dynamics of the transition. Among these, we can mention many. Indeed, the sociology of energy deals with a growing number of social issues that mix with the more technical aspects of the energy sector: the socio-technical dimensions of the transition; the role of businesses and consumers; how consumers change their status, also becoming prosumers; the conflicts that arise in the location of new plants, which rediscover territorial spaces and resources; the revival of social and territorial inequalities, as a result of some types of policies that want to accelerate the transition; the consequences in international relations; and issues related to the relationship between energy and power, resulting from the centralisation and multiplication of producers. A type of study that holds all these perspectives together is the comparative one, which investigates the different transitions at the national level.

It is a line of study that is affected by the analyses of the variety of capitalisms. Following this approach, each model of capitalism would present different models of transition. In any national capitalism, the specific regulation between state, market and society would have created different social trajectories in the transition.

Considering this kind of analysis pertinent, we have chosen to investigate some sociological aspects that we think are relevant in the Italian transition model, a model that allowed the transition to be guided by market logic, with a preponderance however in investments by large multinational companies and public utilities. There are few signs of a bottom up transition through patterns of socialisation of energy production and consumption. Also for this reason, faced with so many projects related to the exploitation of renewable sources, many local committees have arisen in opposition to renewable energy plants. As we analysed in Chap. 3, the reasons for the opposition to the plants cannot be simplified with the Not in My Back Yard (NIMBY) rhetoric. There are a variety of social factors and processes that shape the relationship between renewables and society (Magnani & Osti, 2016), from the cultural aspects that guide the perception and management of risk, to the issues of distributive and procedural justice in relation to the costs and benefits of new plants, to the different epistemologies and knowledge on energy technologies, to the role of trust in institutions, to the territorial rooting of production and consumption networks.

Starting from these reasons, in Chap. 4 we wanted to explore some concrete examples of distributive injustice. We have looked at how energy transition policies impact differently on different social classes and places. According to several authors, the investigation on the impacts of policies also helps to identify the causes of so many hostilities towards the energy transition. There are those who are benefited and harmed by the transition, namely those who see their quality of life improve and those who do not receive positive effects from the changes. Within the category of those harmed, some of the hostilities to new plants for renewable sources should be sought. However, as we argued in Chap. 3, the reasons for the social undesirability of the transition arise from many dimensions. Within this way of looking at the transition, spatial and territorial factors grow in importance. This emerges clearly in the various chapters of the book: from the concept of the backyard to the analysis of social and territorial inequalities, up to the localisation of some experiences of energy socialisation. Considerable sociological literature has focused on the environmental and location conflicts of new plants. Little attention, on the other hand, has been paid to the so-called non-oppositional collective actions. Chapter 5 focuses on this type of mobilisation. New forms of sociality are also emerging around renewables. They have been variously interpreted by the literature as grassroots innovations, social enterprises, strategic fields of action for social movements, socio-ecological systems and socio-technical assemblages. These concepts identify several relevant social factors that mediate the relationship with energy: from relational social resources, such as social capital and mutuality, to symbolic resources linked to the framing of issues and the creation and mobilisation of collective identities, up to the processes of intermediation by ecopreneurs and cooperative organisations.

Recently the phenomenon of energy cooperatives also seems to reflect the emergence of different forms of mobilisation, of a neo-materialist type (Pellizzoni, 2021). The latter assume that change in daily practices is critical in bringing about change at broader levels. Consequently, the development of forms of intermediation aimed at reconstructing a link between people, objects, technological artefacts and places becomes central and is characterised as attempts to replace the market and the state in providing goods and services. However, these experiences remain a minority in the Italian context, as they have problems related to replicability and organisation over large areas.

A task for the social sciences in the coming years will certainly be to investigate even more deeply the social dimensions of the transition, working so that those experiences that socialise energy in the direction of strengthening communities and emancipating people and places can spread more forcefully.

The reconstruction of links not only at the level of sociological discourse but also of empirical practices between social actors, forms of participation and the themes identified here appears to be a very crucial objective for addressing the energy question in a non-superficial way.

References

Carrosio, G., & Scotti, I. (2019). The 'patchy' spread of renewables: A socio-territorial perspective on the energy transition process. *Energy Policy, 129*, 684–692.

Magnani, N., & Osti, G. (2016). Does civil society matter? Challenges and strategies of grassroots initiatives in Italy's energy transition. *Energy Research & Social Science, 13*, 148–157.

Pellizzoni, L. (2021). Nature, limits and form-of-life. *Environmental Politics, 30*(1–2), 81–99.

Index

N. Magnani, G. Carrosio, *Understanding the Energy Transition*,
https://doi.org/10.1007/978-3-030-83481-4

The manufacturer's authorised representative in the EU is Springer Nature Customer Service Centre GmbH, Europaplatz 3, 69115 Heidelberg, Germany. If you have any concerns regarding our products, please contact ProductSafety@springernature.com

Printed and bound by CPI Group (UK) Ltd, Croydon, CR0 4YY
30/09/2026
02225711-0001